建筑立场系列丛书 No.32

居住的流变
Dwelling Shift

中文版

韩国C3出版公社 | 编
张琳娜 于风军 耿婷婷 | 译

大连理工大学出版社

004 智能街道设施 _ JunSang You
008 新加坡科学设计大学的图书馆 _ City Form Lab
012 黑色草莓树太阳能充电设施 _ Miloš Milivojević
013 围绕在自然身边 _ St-André-Lang Architectes
016 加油站和麦当劳店 _ Giorgi Khamaladze
020 港口新城大学地铁站 _ Raupach Architekten
022 过山车式建筑 _ Atelier Zündel Cristea

居住的流变

024 *我们的住宅：城市住宅和现代住宅范例的流变* _ Jorge Alberto Mejia Hernández
030 洛尔蒙住宅 _ Habiter Autrement + Ateliers Jean Nouvel
038 Icod的12所住宅 _ DAO
046 San Vicente Del Raspeig的32所住宅 _ Alfredo Payá Benedito
056 俄罗斯方块：社会住宅和艺术家工作室 _ Jaques Moussafir Architectes
064 巴黎的30座社会住宅 _ KOZ Architectes
072 圣玛丽亚住宅 _ Hierve
084 Plantoun的小村庄 _ Agence Bernard Bühler
094 44个社会住宅单元 _ LEM+ Architectes
100 中心村落 _ 5468796 Architecture + Cohlmeyer Architecture Limited
108 红苹果住宅公寓 _ Aedes Studio

服务于城市

118 *服务于城市* _ Aldo Vanini
124 萨拉曼卡市政厅 _ Carreño Sartori Arquitectos
134 韦克斯福德郡议会总部 _ Robin Lee Architecture
142 毕尔巴鄂市政厅 _ IMB Arquitectos
152 萨莫拉郡议会建筑 _ G+F Arquitectos
160 莱希河畔兰茨贝格市政厅 _ Bembé Dellinger Architekten
166 巴埃萨市政厅的修复 _ Viar Estudio Arquitectura
178 Archidona的文化中心和市政厅 _ Ramón Fernández-Alonso Borrajo

建筑立场系列丛书 No.32

004 Smart Street Furniture _ JunSang You

008 SUTD Library Pavilion _ City Form Lab

012 Black Strawberry Tree Solar Charger _ Miloš Milivojević

013 Turn around the Nature _ St-André-Lang Architectes

016 Fuel Station and McDonalds _ Giorgi Khamaladze

020 HafenCity University Subway Station _ Raupach Architekten

022 Architectural Ride _ Atelier Zündel Cristea

Dwelling Shift

024 *Our House: Urban Dwelling and the Contemporary Paradigm-shift*
    _ Jorge Alberto Mejia Hernández

030 Lormont Housing _ Habiter Autrement + Ateliers Jean Nouvel

038 12 Houses in Icod _ DAO

046 32 Housing in San Vicente del Raspeig _ Alfredo Payá Benedito

056 Tetris, Social Housing and Artist Studios _ Jaques Moussafir Architectes

064 30 Social Housing in Paris _ KOZ Architectes

072 Santa Maria Housing _ Hierve

084 Small Village, Plantoun _ Agence Bernard Bühler

094 44 Social Housing Units _ LEM+ Architectes

100 Center Village _ 5468796 Architecture + Cohlmeyer Architecture Limited

108 Red Apple Apartments _ Aedes Studio

Serving the City

118 *Serving the City _ Aldo Vanini*

124 Salamanca City Hall _ Carreño Sartori Arquitectos

134 Wexford County Council Headquarters _ Robin Lee Architecture

142 Bilbao City Hall _ IMB Arquitectos

152 Zamora County Council _ G+F Arquitectos

160 Landsberg am Lech Town Hall _ Bembé Dellinger Architekten

166 Baeza Town Hall Rehabilitation _ Viar Estudio Arquitectura

178 Cultural Center and the New City Hall of Archidona _ Ramón Fernández-Alonso Borrajo

街道设施 Street Furniture

智能街道设施 _JunSang You

一座城市是不同需求交织在一起的地方。当一条长凳可能成为一处障碍,抑或白天路灯没有必要存在时,公共设施的必要性便会依据城市环境和时间的变化,而变得相关。对于近日的城市景观来说,太多的公共设施混乱地交织在一起,在限定的空间内满足所有不断变化的个人要求。这是另外一种公共设施方面的浪费,它们占领了人们不需要的街道的大部分面积。

以智能手机为代表的融合技术是一个重要的关键词,它不仅仅能够治理凌乱的城市景观,同时也能与其他传统的设计手法区分开来。当大范围的公共设施因为都采用统一的设计而结合起来时,这些连为一体的设施将会被更广范围的使用者(而非所有个体)有效地利用,因为它们使空间和资源得到了最好的利用。

在这一前提下,建筑师规划了一种智能街道设施,将不同的功能性设施,如长凳、街灯、太阳能充电设施、自行车架以及遮阳装置连为一体。

来自大自然的灵感

该项目的设计灵感直接来自于自然,如向日葵、百合花以及伞形结构。

最大效率的太阳能光电

从技术上讲,固定的太阳能平板系统不能产生最多的太阳能。要使光伏系统获得最大的效率,嵌板必须追踪太阳的运行轨迹,以和光线呈垂直状态。如果光伏系统带有太阳轨迹追踪系统,那么它便使平板系统的光伏效率提高30%~50%。

提高现存遮阳系统的效率

现存遮阳系统的最大问题是相对于低角度的阳光来说,其遮阳效率非常低。然而,要覆盖住全部的座位区是十分不容易做到的,因为太阳光的角度随着季节和时间的变化而变化,且变化十分明显。

嵌板与太阳成垂直状态,使其能够在同一时间投下最大的阴影。这一设施作为一个充电设施和遮阳设施来说,通过安装太阳轨迹追踪系统,能够产生了最大的功效。

Smart Street Furniture

A city is a place where different desires are mixed. As a bench can be an obstacle or there's no need for a streetlamp during the day, the necessity of public facilities are quite relative depending on the urban circumstances and time. It has been disorderly entangled by too many different public facilities for a cityscape today while satisfying all of the ever-changing individual needs of people within confined spaces. It is just another waste for the public facilities occupying a large portion of a street to people who don't use them.

The fusion technology represented by the

项目名称：Smart Street Furniture
建筑师：JunSang You
甲方：Modern Atlanta
用途：multifunctional public facility
结构：TWIP steel
材料：solar cells, TWIP steel, aluminum, LED lights
造价：USD 8,000 (prototype cost)
设计时间：2013

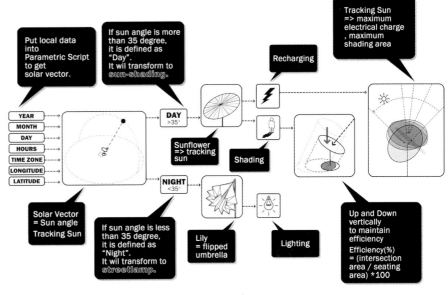

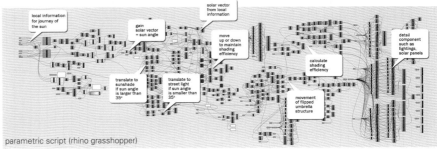

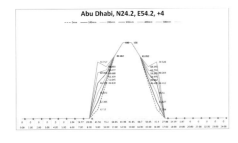

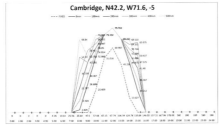

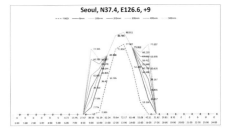

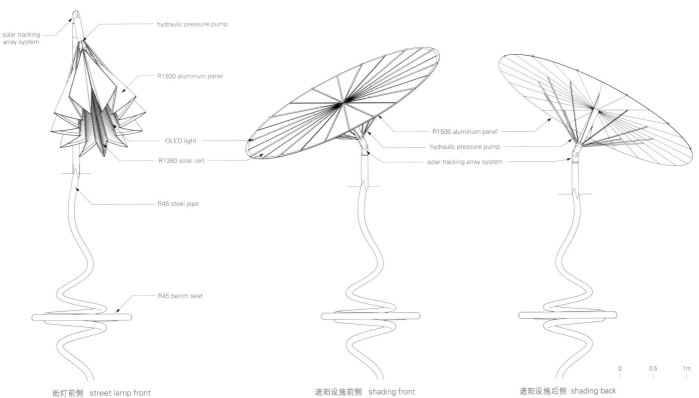

街灯前侧 street lamp front　　遮阳设施前侧 shading front　　遮阳设施后侧 shading back

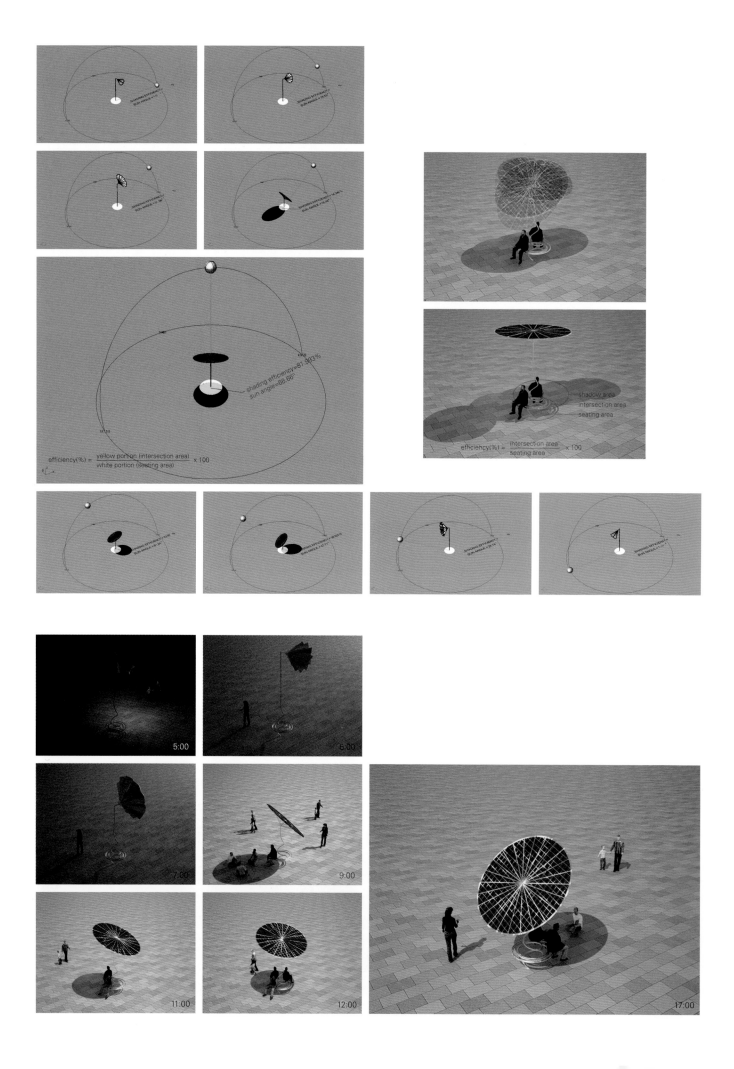

smart phone is an important key, not only to cure the messy cityscape but also to differentiate from other conventional design methodologies. When a broad range of public facilities are combined into one convergent design, the united facility will be efficiently used by a much wider range of users than all individuals because it makes the best use of the space and resources.

On this premise, I proposed a new Smart Street Furniture combining different functional facilities such as a bench, streetlamp, solar recharger, bike rack, and sun-shading device into one item.

Inspiration from Nature

The design of this project is directly inspired by nature, such as sunflower, lily, and umbrella.

Maximum Efficiency for Photovoltaics

Technically, the fixed flat plate solar system cannot generate the maximum solar energy. To get maximum efficiency of PVs, the panel has to track the motion of the sun to be perpendicular to sunlight. If PVs has the Solar Tracking Array System, it would increase the photovoltaic efficiency of the Flat Plate System by 30~50 percent.

Improvement of Efficiency of Existing Sun-Shading System

The biggest problem of the existing sun-shading system is that it is especially vulnerable to the low angle of the sunlight. It is not easy to make it large enough to cover the whole seating area because the angle of sunlight varies dramatically with the season as well as the time.

The panel facing perpendicularly to the sun would cast the biggest shadow at the time. By just installing the solar tracking

system, this device would achieve the best result as a recharger and shading device.

JunSang You

新加坡科学设计大学的图书馆 _City Form Lab

新加坡科学设计大学的图书馆坐落临时的多佛校园的一个斜坡草坪上。

该图书馆周围有三棵成熟的大树,以隔离来自北面Ayer Raja高速公路的噪音,其网格式外壳结构充分利用了受到限制的场地,并且激活了位于现存图书馆建筑后身的室外空间。在白天,它提供了一处阴凉的开放空间,供人们休闲、工作以及与学校师生交往之用。在夜晚,它则成为一处非正式的集会、夜晚补习以及学校社交活动的场所。工作桌、可移动的书架以及无线网络将这座图书馆转换为一处宿舍和教室之间的"第三空间",在这里,人们可以在轻松的氛围内来进行知识交流,以及社交活动。

图书馆的天篷构建于一个长形链状的传统结构之上,这个结构仅仅使用了少许的材料便能产生大面积的跨度,天篷形成了一个轻质的木材外壳,没有支柱、梁和直墙。设计采用了一个悬垂链式的数字化模型,以形成一个有效的双曲面外形,这个外形与受压所产生的线形推力保持一致。

电脑设计和电脑控制的建造能够使图书馆形成复杂的三维形式,且使用了最小的成本,采用了易获得的材料以及流线形的组装过程。与钢质的网格外壳不同,它没有复杂的三维结构节点,所有构件都是由在新加坡使用计算机数控制成的、严格把关的扁平胶合板以及预制钢板建造的。因此,现场的工作则包括有序地组装3000块独特的胶合板以及600块独特的金属薄板瓦,这些基于一种绘图——一种三维拼图的数字地图,能够显示哪块与哪块相连接。

在切割的过程中,每块胶合板和覆层构件都刻上了ID号,并且在完成的结构上清晰可见,以作为一处装饰。一年级的学生在预先组装构件的过程中给予了帮助,而承包商则在现场进行结构安装。

这座图书馆在两年之后可以拆卸,并且可回收。

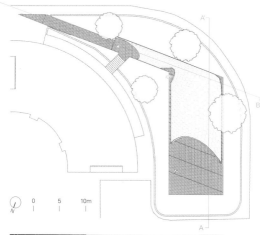

SUTD Library Pavilion

The Singapore University of Technology and Design(SUTD) library pavilion is located on a sloping lawn on the temporary Dover Campus.

Accommodating three mature trees and forming a noise barrier toward the Ayer Raja Expressway in the north, the gridshell structure of the pavilion harnesses the site constraints and activates an outdoor space behind the existing library building. During the day it offers a shaded open-air place to relax, work, and mingle for students and staff of the university. At night it becomes a place for informal gatherings, evening lectures and SUTD community

项目名称：SUTD Library Pavilion
地点：Dover Rd, Singapore
建筑师：City Form Lab
工程师：ARUP
施工：Arina International Hogan(AIH),
SUTD students, staff
面积：covered_200m², deck_300m²
材料：plywood panel, plywood block, steel cladding tile
竣工时间：2013.5
摄影师：
Courtesy of the architect-p.8, p.10
©Philipp Aldrup (courtesy of the architect)-p.9

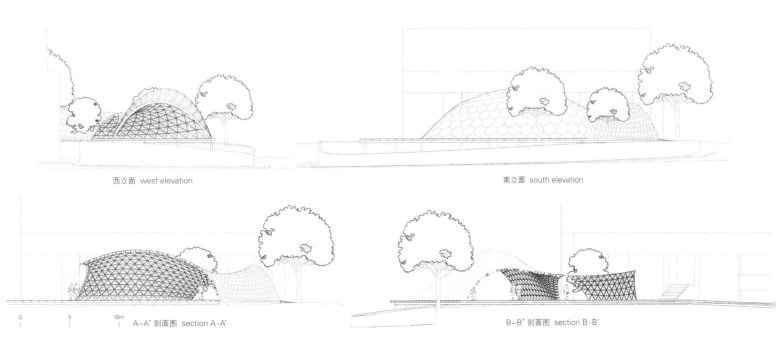

西立面 west elevation   南立面 south elevation

A-A' 剖面图 section A-A'   B-B' 剖面图 section B-B'

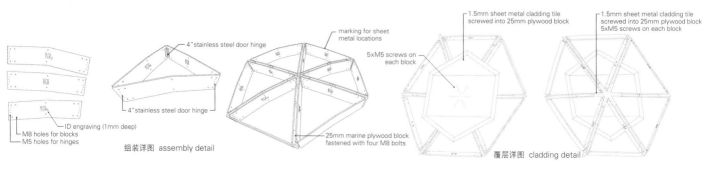

events. Work-desks, mobile bookshelves and wireless Internet transform it into a "third space" between the dormitory and the classroom where intellectual and social exchanges occur in a casual atmosphere. Building upon a long tradition of catenary structures that use little material to achieve considerable spans, the canopy forms a lightweight timber shell with no columns, beams, or vertical walls. A numeric hanging-chain model was used to determine an efficient double-curvature shape that follows the lines of thrust in compression.

Using computational design and computer controlled fabrication allowed the pavilion's complex three-dimensional form to be achieved with readily available materials and a streamlined assembly process at minimal cost. Unlike steel gridshells, it has no complex three-dimensional structural joints – all of its elements were prefabricated from strictly flat plywood and galvanized steel sheets on CNC machines in Singapore. The site work thus comprised an orderly assembly of 3,000 unique plywood and 600 unique sheet-metal tiles based on only one drawing – the numeric map of a three-dimensional puzzle indicating which piece fits next to other pieces.

ID numbers were engraved in the cutting process on each plywood and cladding element, which remain visible in the finished structure as ornament. First year SUTD students assisted with the pre-assembly of the pieces and the contractor erected the structure on site.

The pavilion is designed to be dismantled and recycled after two years.

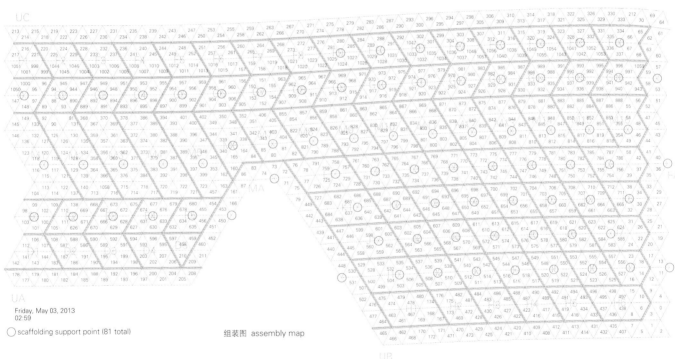

组装图 assembly map

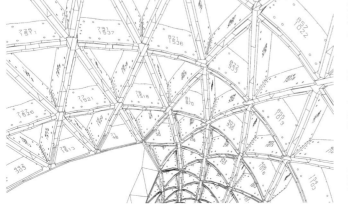
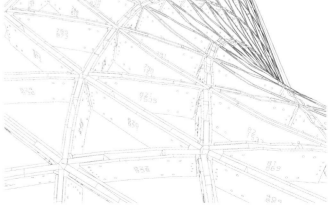

黑色草莓树太阳能充电设施 _ Miloš Milivojević

塞尔维亚建筑师Miloš Milivojević发明了一种全新的公共充电设施设计，可以为草莓能源公司发明的移动设备进行充电。这一太阳能系统利用太阳能，能使Tašmajdan公园里面的游客为他们的移动手机、平板电脑以及多媒体设备的电池充电。

这棵黑色的草莓树是一棵人造树，能够将太阳能转化为电能，从而与周围的树木融为一体，共同为地球获得更多的氧气而奋斗。一棵真正的树吸收二氧化碳，释放氧气，使人们的环境更加清洁，而这棵黑色草莓树则利用太阳能，生产无害物质，这种方法致力于全球二氧化碳的减排，使Tašmajdan公园更加环保。

黑色的草莓树便于用户使用，其构造较为纤细，采用人工来构造，使其本身看起来如同雕塑一般。这一大型且优雅的钢结构的长度超过了3.5m，高度超过了4.5m，与一棵真正的树的线条保持一致。太阳能充电设备上方的方形表面采用9块薄膜玻璃制成的太阳能电池板以及9块玻璃支承嵌板来覆盖，它们同时还形成一个屋顶，在恶劣的天气条件中起到保护的作用。这一太阳能充电结构是由人工树顶托起来的，树顶包括8条钢质树枝。

木质长凳的长度超过了4m，位于黑色草莓树的前方，这一大型长凳可以容纳很多人。充电接口位于悬挂在长凳的金属杆上的可伸缩的绳子上。该设备功能方面所使用的全部必要技术都安置在一个钢质盒子中，隐藏在木凳之下。因其拥有这一功能，这棵树成为提醒人们未完全挖掘太阳能潜力的一个永久性存在物。

Black Strawberry Tree Solar Charger

Serbian architect Miloš Milivojević developed completely a new design for public solar charger for mobile devices invented by Strawberry Energy Company. This solar system enables visitors of Tašmajdan Park to recharge the batteries of their mobile phones, tablets and multimedia devices with the energy of sun.

The Strawberry Tree Black is conceived as an artificial tree which transforms solar energy into electricity, thereby joining the surrounding trees in a common struggle for the planet richer in oxygen. As real trees absorb CO_2 and release oxygen which makes our environment cleaner, the Strawberry Tree Black uses clean solar energy and produces no detrimental substances, and this way also contributes to global CO_2 emission reduction and makes Tašmajdan Park greener.

Strawberry Tree Black is user-friendly with the thin and artistic construction which makes it look like a sculpture. The large but elegant steel construction is more than three and a half meters long and four and a half meters tall which follows the line of a real tree. The square surface at the top of the solar charger is covered with nine thin-film glass solar panels and nine supporting glass panels which simultaneously acts as a roof, protecting against bad weather conditions. This solar cell structure is held by the artificial treetop consisted of eight steel branches.

The wooden bench, more than four meters long, is positioned in front the Strawberry Tree Black and large enough to host a lot of people. Charging points on stretchy cords hang from the metal bar on the bench. All necessary technique for the functioning of the device is placed in a steel box hidden in a wooden bench. With its function, this Strawberry Tree Black acts as a constant reminder of the insufficiently exploited potential of the Sun's energy.

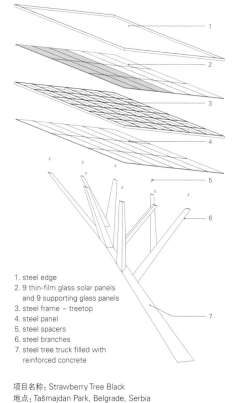

1. steel edge
2. 9 thin-film glass solar panels and 9 supporting glass panels
3. steel frame – treetop
4. steel panel
5. steel spacers
6. steel branches
7. steel tree truck filled with reinforced concrete

项目名称：Strawberry Tree Black
地点：Tašmajdan Park, Belgrade, Serbia
建筑师：Miloš Milivojević
结构工程师：Milan Zlatanović
甲方：Strawberry Energy Company
竣工时间：2012.11

无限循环 Cycle Circle

围绕在自然身边 _ St-André-Lang Architectes

身为Archi<20竞赛的获奖项目，且有7000欧元的预算资金，这个20m²的住宅原型在法国东北部Muttersholtz村的一处受保护的区域建立起来。这座小亭子——其设计遵循了一个简单的生成形状，因此命名为循环——能够对周围所有的景观进行感知。

建筑中间设置了一口采光井，以形成特色，室内布局也因此根据太阳的位置以及其日常的循环，即从东到西，来进行选择。一个木质家具沿着建筑边界进行拓展，将不同的日常活动需求完美地结合在一起。在北侧——入口一侧——一处低矮的区域（夜晚的休息空间）通向建筑东部的工作空间，以及南部一处更宽敞的区域，这处更宽敞的区域面向天空开放，在立面，洞口的设置也完全取决于太阳的方位：事实上，设计与所有的自然元素都是紧密相连的。

立面，除了作为室内外之间的屏障之外，也变得具有功能性，并且呈现出了一个全新的规模：受到了阿尔萨斯平原的玉米烘干机的启发。立面随着季节的变化而变化，它模糊了对物体的即时解读，几乎抹掉了房子的痕迹。

中央庭院位于居住空间的核心，反映了让自然进入房屋的愿望。建筑周围种有本地植物，成为阿尔萨斯里德自然景观的一个象征；通过这种方式，居住空间位于未受破坏的自然与受到控制的栽培区域之间的交汇点，象征着最终的平衡，称为可持续发展的基础。

Turning around the Nature

Winning project of the Archi<20 competition's selection and with a 7,000 Euro budget, this 20m² housing prototype has been constructed in a protected natural area in the village of Muttersholtz, in the northeast of France. The pavilion – designed following a simple generative shape, namely a circle – allows the perception of the entire surrounding landscape.

Characterized by the presence of a light shaft in the middle of the building, the indoor set-up has consequently been chosen according to the Sun's position and its daily cycle, that is from East to West. The

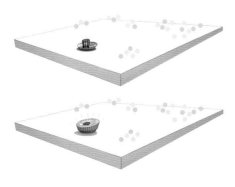
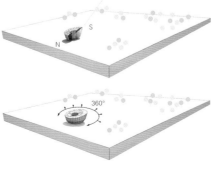
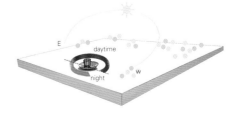

furniture, consisting of just one block extending around the entire house, perfectly integrates the needs of the different daily activities. On the Northern side – the entrance side – a low-ceiling area (night space) leads to a working one in the eastern part of the building and to a more generous space in the southern part, opening up to the sky. On the facade, the rhythm of the openings depends on the Sun's position as well: as a matter of fact, the design is closely linked to all the natural elements. The facade, beside operating as the partition between the inside and the outside, becomes functional and takes on a new dimension: inspired by the corn dryers in the Alsatian plains. Changing along with the seasons, the facade blurs the immediate reading of the object almost erasing the house itself.

The central patio, at the heart of the living space, mirrors the desire to let nature enter the house. Planted with local species, this building becomes a metaphor for the natural landscape of the Alsatian Ried: in this way, the living space lies at the meeting point between unspoilt nature and controlled cultivation, becoming the symbol of a finally found balance and the basis for sustainable development.

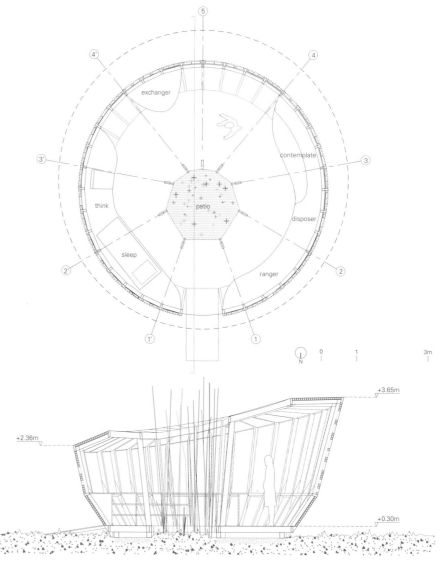

项目名称：Tourner autour du Ried
建筑师：Bastien Saint-André, Maxime Lang
地点：Muttersholtz, France
总建筑面积：20m²
材料：douglas fir, stainless steel wire netting, corn cobs, acrylic panels, cardboard columns, PVC free canvas
造价：EUR 7,000　竣工时间：2012.5

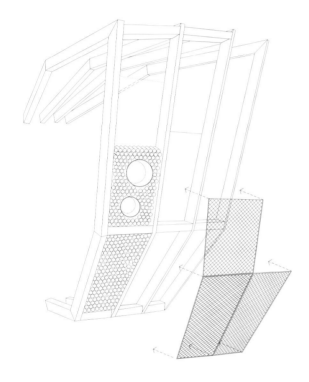

隔离式的一体化 Isolated Join

加油站和麦当劳店 _Giorgi Khamaladze

该项目位于格鲁西亚的海滨城市巴统的一个新建的城市化区域。它包括加油站、麦当劳店、休闲空间以及反射水池。

项目位于场地的中心位置,十分重要,因此,建筑师决定通过限制建筑的占地面积以及机动车路线的方式,尽可能大地将休闲区域返回给城市。因此,所有的功能都压缩在一个体量内。

空间构成是这样的,主要的两个功能区——机动车服务区和餐饮区从地理位置和视觉上都相互隔离,使餐馆用餐的顾客看不见加油站的所有操作过程。

由于预先定义的小型建筑面积,大部分的支承空间和效用空间均加以分组,且位于一层,以和所有的技术接入点相接近。

餐馆的公共空间始于大堂,其专门的入口位于一层。从那里看,作为一种自然连接上层且为顾客提供楼层之间的平缓过渡体验的方式,楼梯呈上升趋势,并且在中间层创造了可居住的平台,作为餐饮空间。

部分餐饮空间提供了望向室外水景的视野,同时其他部分完美地过渡到开放的上层室外庭院内。庭院四周均围合起来,使空间免受室外噪音的打扰,提供了一处安静的室外座位区。植被层覆盖了整个加油站巨大的悬挑天篷,增添了自然氛围,且对于露台来说,成为一处生态防护层。

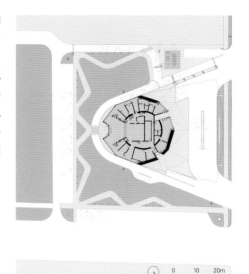

项目名称: Fuel Station + McDonalds
地点: Batumi, Georgia
建筑师: Giorgi Khmaladze
结构工程师: Capiteli MEP工程师: Gulfstream
MCD标准: Archange & Schloffer
厨房设计工程师: Franke 室外照明师: Erco
甲方: SOCAR
用地面积: 5,000m² 总建筑面积: 1,200m²
设计时间: 2010~2011 施工时间: 2012~2013

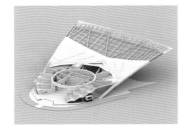
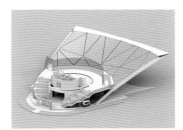

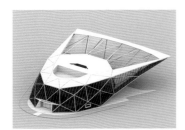

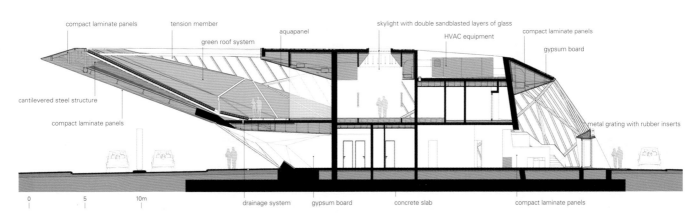

A–A' 剖面图 section A-A'

Fuel Station and McDonalds

The project is located in one of the newly urbanized parts of the seaside city of Batumi, Georgia. It includes fuels station, McDonalds, recreational spaces and reflective pool.

Given the central location and therefore importance of the site, it was decided to give back as much area as possible for recreation to the city by limiting the footprint of the building and vehicular circulation. This resulted in one volume with all programs compressed within.

Spaces are composed in such way, that two major programs – vehicle services and dining are isolated from one another, both physically and visually so that all operations of fuel station are hidden from the view of the customers of the restaurant.

Because of the predefined, small building footprint, most of the supporting and utility spaces are grouped and located on the ground level to be close to all technical access points.

Public space of the restaurant starts from the lobby and its dedicated entrance is on the ground floor. From there, as a way to naturally connect to the upper floor and to offer customers the experience of smooth transition between levels, the floor steps upwards and creates inhabitable decks on intermediate levels to be occupied as dining spaces.

Part of the dining space offers views towards outside water features, while the rest of it seamlessly transitions into open air patio on the upper level. The patio, enclosed from all sides to protect the space from outside noise, provides calm open air seating. The vegetation layer, which covers the cantilevered giant canopy of the fuel station adds natural environment and acts as a "ecological shield" for the terrace.

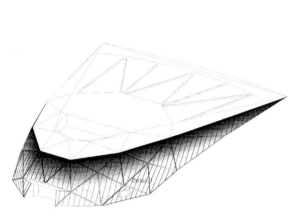

陶瓷熔块形式的结构 ceramic frit pattern scheme

1. insulated facade glazing system
2. structural profile 80x120mm
3. 30mm thermal insulation layer structural steel profile R140mm
4. structural steel profile 120mm
5. outer layer 8mm low-e coated
6. inner layer 6mm lamiglass with ceramic frit custom pattern
7. folded aluminum composite panel

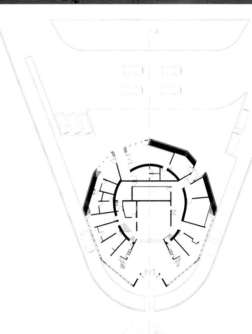

一层 first floor

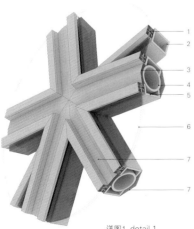

详图1 detail 1

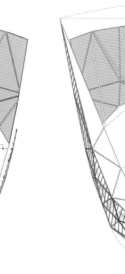

二层 second floor

屋顶 roof

光容器 Container Light

港口新城大学地铁站 _Raupach Architekten

这个设计对场地的特征产生了一定的影响：由于光线、钢质外壳以及运输容器模块的作用，砖立面自身的面貌发生了改变，其颜色随着季节的变化也发生了改变。

因此，地铁站点的材料主要以钢、光线以及颜色为主导。

光线被抓进容器中。建造一座充满自信和力量的物体（地铁站），使其与钢质材料共存。地铁站与运输容器之间的正式连接变得清晰，其设计理念是一处海洋环境。

12个光容器盘旋在站台的上方，控制着10m高、16m宽的站台区域，对其氛围起决定性作用，并且赋予了站台力量和不同。

彩色的映像都严格地映射在了闪闪发光的黑色钢板上，钢板固定在站台的墙体和天花板表面上。

因为它们有静态和动态两种程序选项，因此，它们也可与一天中情绪的变化保持一致，也可为等候列车的旅客带来一种感官体验。

所有的容器都是半透明玻璃制成的，每个容器为6.5m长，2.8m高，立面配有280RGB的LED灯。其底部采用均匀照明，以在站台上投下一束温暖且均匀的灯光，与规划的光线区/黑暗区形成对比。

射灯设备位于天花板，为柜台区域营造了舒适的照明氛围。

HafenCity University Subway Station

The design reacts associatively on the located identities on site: the brick facades change their appearance due to daylight, the steel hulls and modules of transport containers, and change their colors with the seasons.

Consequently, the materiality of the subway stop is guided by Steel, Light and Color.

The light is grasped into containers. Creating self-confident, powerful objects,

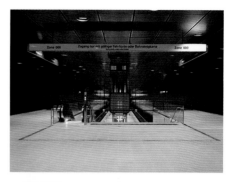

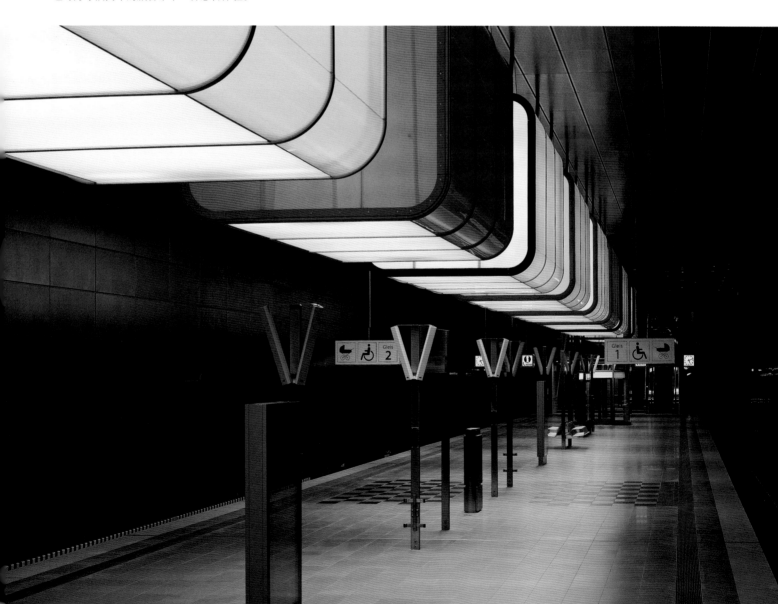

which are able to coexist with steel. Its formal connection to a transport container becomes clear, and its idea is a maritime environment.

Twelve "Light-Containers" hovering above the platform dominate the 10m high and 16m wide platform area, determine its atmosphere, and give creative power and distinctiveness.

Colored Reflections are discreetly mapped onto the shimmering dark steel plates clamped on walls and ceiling surfaces of the platform.

Since they are statically or dynamically programmed by choice, they change themselves in harmony with the different moods of a full day and also bring waiting for the trains a sensual experience.

All containers are vitrified semi-transparently, each 6.5m long, 2.8m high and equipped with 280 RGB-LEDs. The bottom side, which is homogenously illuminated, casts a warm and even glow on the platform and supports the planned light/dark contrasts.

Downlights set in the ceiling trace are the reason for a comfortable lighting atmosphere in the counter areas.

项目名称：Subway Stop: U4 HafenCity University
地点：Hafencity, Hamburg, Germany
建筑师：Christian Raupach
项目团队：Andreas Pabst, Stephan Radlingmayr, Christian Groneberg
开发商：Hochbahn AG
施工监理：Agather Bielenberg
照明规划：pfarré lighting design, d-lightvision
技术设计：Stauss Grillmeier
参与竞赛：2005 (winning proposal)
竣工时间：2012.7

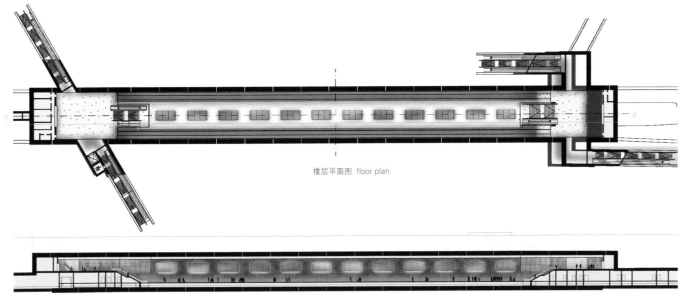

楼层平面图 floor plan

A-A' 剖面图 section A-A'

城市再生 Urban Regeneration

过山车式建筑 _Atelier Zündel Cristea

一座有力量的神殿

由于临近冷却水,巴特西电站因此建成了,场地位于南岸一处61 000m²的地面上。

在初期,该发电站非常受欢迎。它象征着进步、工业以及一种新型力量:人类的力量。在其过去的70多年的历史中,这座发电站展现了其标志性的地位,这在许多形式的流行文化中均有所体现。

一处展现建筑乐趣的新场地

这个项目设想了巴特西一座新公园内场地的再生性,将休闲与建筑结合起来,以创造欢迎所有人的流行场所,致力于心灵和身体的愉悦,使人们享有满满的独特体验。

一座博物馆建筑,以巴黎城内的建筑模型为基础,将通过一系列的画廊来展示从中世纪到现在的建筑与文化遗产的全景。

建筑师试图记住人们将要拜访新巴特西建筑博物馆的主要原因:一次见识和体验建筑、学习建筑并且将其作为一种职业,同时与其他人进行探讨的机会;探索发电站的建筑背景;重新熟悉艺术与建筑作品。

一处心灵与身体体验的场地

建筑师已经将异域的围栏元素引入发电站的空间中,这将首先激活空置的空间。它将引领游客进驻主要路径中,并且使用最小的力量来呈现出建筑的全部布局。因为围栏的位置划定了主要路径,那么在发电站的室内外空间穿梭便开始有意义。这个项目将发电站置在一个中心舞台之上,结构本身通过其令人印象深刻的规模、建筑以及独特的砖材料来对场地进行强化。建筑师建造的路径与一系列的空间相连接,使人们自己去发现博物馆前方的广场、空地,室外、室内以及上方的小径,以及穿过庭院和展览室的小径。

这个项目具有在当今时代唤醒人类维度和尺度的力量,并且质疑人类与结构之间的关系。这并不仅仅是一次展示,还是一首后工业时代的诗歌。

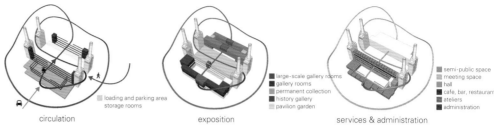

Architectural Ride

A Temple of Power

The Battersea Power Station was built, due to the proximity of the cooling presence of water, on a 61,000m² plot of land situated on the south banks.

From its inception, the station was very popular. It symbolized progress, industry, and a new type of power: the Power of the People. Over its seventy year history, the station has taken on an iconic status, having been represented in many forms of popular culture.

A New Site for Architectural Pleasures

This project envisions the regeneration of the Battersea site within a new park com-

bining leisure and architecture, in creating a popular spot welcoming to all, dedicated to the pleasures of mind and body, replete with unique experiences.

A museum of architecture, based on the Parisian Cité de l'Architecture model, will through a series of galleries present a panorama of architecture and cultural heritage from the Middle Ages to today.

The architects tried to keep in mind the principal reasons why people would visit the new Battersea Museum of Architecture: the opportunity to see and experience architecture while learning about it as a profession and discussing it with others; exploring the architectural setting of the power station; revisiting familiar works of art and architecture.

A Playground for the Mind and the Body

The architects have introduced the foreign element of a rail into the space of the power station, which will function above all in animating the empty space. It will offer visitors entering the structure of a primary pathway, allowing them to take in the essential layout of the building with a minimum of effort. With the pathway determined by the presence of the rail, the simple fact of moving through the exterior and interior spaces of the station begins to make sense. This project puts the power station on center stage, the structure itself enhancing the site through its impressive scale, its architecture, and its unique brick material. The architects created pathway links together a number of spaces for discovery: the square in front of the museum, clearings, footpaths outside and above and inside, footpaths traversing courtyards and exhibition rooms.

The project has the strength of evoking the dimension and scale of man in the contemporary era, putting into question people's relationship to the structure. It is not only a matter of showing, but also of suggesting post-industrial poetry.

Atelier Zündel Cristea

项目名称：Architectural Ride 地点：London, UK
建筑师：Atelier Zündel Cristea
项目团队：Gregoire Zundel, Amilcar Ferreira, Charles Wallon, Tanguy Aumont
甲方：archtriumph 用途：museum of architecture
用地面积：155,000m² 有效楼层面积：43,692m²
结构：building_brick, concrete/roller coaster_steel

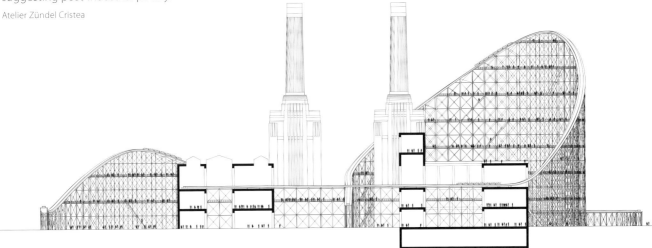

居住的流变

在建筑师Carlo Aymonino于1971年出版的《理性的住宅》一书中，他利用了一个非常有趣的假设，分析了两次国际现代建筑协会所讨论的城市住宅规划，至少直到20世纪70年代，1920年代和1930年代的住宅模型仍然处于应用中，且基本上没有受到更新的或者是更创新的提议的挑战。

今天，新的建造方法倾向于承认建筑师作为一个充满强度和永久性变化的整体的地位。我们不仅仅将建筑认为是以主流现代主义为特色的独白，而且还是一场非正式的演说。从另一方面来说，我们除了将建成环境认为是在一个独立的创造者与其设计之间的随意关系外，还将其看成是一件人工制品，它将集体目标具体化，并且挑战着创造性所控制的理念。

自第二次和第三次国际现代建筑协会举办之后，时代必然发生了变化。但是，那些包含大面积的建成环境的住宅规划是怎样反映这一变化的呢。

一系列的（10个）城市住宅项目通过建筑最简单的工具完成的实际解决方案来进行解析，可以被解读为我们这个时代建造建筑所用原则的缩影，并且为我们在将来提供了有力的洞察力，在对建筑范例进行重新评估时，一种决定性的转变允许专业性讨论的出现，以进行未知的且非常有用的探讨，同时对属于我们这个时代的Aymonino于20世纪70年代提出的宣言的有效性进行检验。

In his 1971 book *The Rational Dwelling* the architect Carlo Aymonino analyzed the urban housing programs discussed at two CIAM meetings, and came up with an extremely interesting hypothesis. Until the early 1970s at least, the residential models of the 1920s and 30s remained in full use and were basically unchallenged by newer or more innovative propositions.

Today, new approaches tend to acknowledge architecture's status as one full of intensity and permanent change. More than the monologues that characterized mainstream modernism, we can now speak of architecture as an informal discourse. On the other hand, beyond the causal relation between an individual creator and his design, we can speak of the built environment as an artifact that materializes collective aims, challenging the idea of creative control.

Times have certainly changed since the second and third CIAM meetings. But, how has the dwelling program that constitutes the vast majority of the built environment reflected this change?

Dissected into practical solutions within architecture's simplest tools, a selection of ten urban housing projects can be read as the epitome of the principles that generate architecture in our time, and thus provide us with useful insight into the future. A critical shift in the reassessment of architectural paradigms should allow the professional debate to delve into uncharted and very promising discussions, while testing the validity of Aymonino's claims of the seventies in our time.

Dwelling Shift

洛尔蒙住宅/Habiter Autrement + Ateliers Jean Nouvel
Icod的12所住宅/DAO
San Vicente Del Raspeig的32所住宅/Alfredo Payá Benedito
俄罗斯方块：社会住宅和艺术家工作室/Jaques Moussafir Architectes
巴黎的30座社会住宅/KOZ Architectes
圣玛丽亚住宅/Hierve
Plantoun的小村庄/Agence Bernard Bühler
44个社会住宅单元/LEM+ Architectes
中心村落/5468796 Architecture + Cohlmeyer Architecture Limited
红苹果住宅公寓/Aedes Studio

我们的住宅：城市住宅和现代住宅范例的流变
/Jorge Alberto Mejia Hernández

Lormont Housing/Habiter Autrement + Ateliers Jean Nouvel
12 Houses in Icod/DAO
32 Housing in San Vicente del Raspeig/Alfredo Payá Benedito
Tetris, Social Housing and Artist Studios/Jaques Moussafir Architectes
30 Social Housing in Paris/KOZ Architectes
Santa Maria Housing/Hierve
Small Village, Plantoun/Agence Bernard Bühler
44 Social Housing Units/LEM+ Architectes
Center Village/5468796 Architecture + Cohlmeyer Architecture Limited
Red Apple Apartments/Aedes Studio

Our House: Urban Dwelling and the Contemporary Paradigm-shift
/Jorge Alberto Mejia Hernández

我们的住宅
城市住宅和现代住宅范例的流变

维也纳、柏林、苏联……在现代主义热情的制高点，效率和合理化达到了近代史上另一个转折点。衰败的"波将金城市"[1]（是过去不健康的、不道德的以及低效的反映）被一种理论框架所取代，在这个框架内，建筑探索所产生的舒适性变得有条不紊。合理化的住宅，即可居住的机器，都被包含在一代人的梦想之中。

利用大量的社会住宅项目来占领土地，这种新颖的方式暗含了政治方面的进步（至少对于马克思主义的拥护者是如此），而住宅本身的形式则在科学严谨性与道德性的光环下被人们细心地钻研。人们依靠于他们的建筑师，而建筑师通过制造他们认为是一次真正变革的全部设备来交付给人们。他们说，一座真正的建筑已经出现了，它包含全新的建筑技术、全新的形式和风格、全新的用途和全新的代表性方法、教育重点以及描述性技术。所有留下来的都是为了人类居住在其间而服务，使人们幸福永远。

不幸的是，他们没有做到。之后数以百万计的居住在智能住宅中的用户自杀了，半个世纪之后，当Carlo Aymonino[2]对于"生存最低限"的方法进行称赞时，国际现代建筑协会具有远见性的宣言则又一次地作为相关数据被发表出来，大量的建于这样的原则基础上的城市住宅项目在一次冲击波中全部消失[3]。

微妙的矛盾

在这些天，评估一套当代住宅项目意味着要深入探讨矛盾。我们的根基在过去进入到了不同的地层，在20世纪20年代和70年代都为我们的今天保留住了珍贵的信息。一方面，Aymonino潜在的声明遵循了在使用不只一种方式的原则下，合理的住宅带有清晰划分的私人区、服务区以及社交区，且集体使用的设备仍然优胜于一切的原则。从另一方面来说，几十年的冲击、缺失以及无数试图超越现代主义的界限来取得进步（带有不等的一系列成果展示）的人和物，定义了我

Our House:
Urban Dwelling and the Contemporary Paradigm-shift

Vienna, Berlin, the Soviet Union… At the acme of modernist enthusiasm, efficiency and rationalization reached yet another point of inflection in recent history. The decadent "Potemkin City"[1] (unhealthy, immoral and inefficient reflection of the past) was superseded by a tense theoretical framework, in which architecture's explorations comfortably fell into place. The rational dwelling, that *inhabitable machine*, embodied the dream of a generation.

Novel ways of occupying the territory with enormous social housing projects implied political progress (at least for the Marxist observer), while the forms of the houses themselves were meticulously studied under the aura of scientific rigor and morality. Humanity relied on its architects, and they certainly delivered by bringing forth the full apparatus of what they believed to be a true revolution. A new architecture, they said, had appeared, encompassing new building techniques, new forms and styles, new approaches to the use and a brand new set of representational methods, pedagogic emphases and descriptive techniques. All that remained was for people to live in their houses, happily ever-after.

Unfortunately, they didn't. The inhabitants of the smartest homes ever built killed themselves by the millions, soon after. Half a century later, while Carlo Aymonino[2] lauded the "Existenzminimum" approach, and the visionary manifestos of CIAM congressmen were once again brought forth as relevant data, vast urban housing projects founded on such tenets vanished with a blast[3].

The contradiction is exquisite.

Assessing a set of contemporary residential projects these days implies delving into this contradiction. Our roots tap into very different strata in our past, and both the nineteen-twenties and the nineteen-seventies hold precious information about the present for us. On the one hand Aymonino's underlying statement abides: the rational dwelling, with its clearly divided private, service and social zones and collective equipments remains unsurpassed, in more than one way. On the other hand, decades of hit and miss and a myriad attempts to progress beyond the limits of modernism (with an uneven set of achievements to show) define the spirit of our time. At a critical level, heretofore it remained possible to

红苹果住宅公寓的规模较大,且带有足够的过渡空间,植被丰富
Red Apple Apartments, large with enough space in-between and plenty of greenery

Plantoun的小村庄,利用绝对有限的建筑用料来达到绝对的利用效率
Small Village of Plantoun, attempting to achieve absolute efficiency in the use of an extremely limited material repertoire

们这个时代的精神。在临界层次,在相关的隔离区域来观察最复杂的现象,或者将其作为统一规划的一部分,仍然是有可能的。但是现在我们不能这样做,至少在符号学的开放性以及系统理论出现之后,至少在互联网的发明以及全球性文化所产生的极端复杂网络的条件下,我们不能这样做。在这些天内,我们的评估不得不具有关联性。

在这一点上,在其他必要的输入中,今天的城市住宅项目,能够被现代主义产物所供养,人们无需去描述个人作品的优点和缺陷,而是阐述一些条例,在这些条例中,每个新项目似乎都对其先例的标准进行争论,并且成为一个更大的、更具有进步性趋势的一部分。但是问题仍然遗留下来:我们在这四十年里,在与法兰克福和布鲁塞尔协议的关系中,我们是在前行吗?如果是这样的话,我们怎样在不与其他立场对立的情况下,来判断我们自己的立场?

一旦一部分被观察的环节加以分析,那么这种进步就会变得有形,而将这些环节与其他实体连接的原则也会被理解。幸运的是,建筑仍然成为一种基本知识的来源。因此,它在一个有限问题的范围内来探索自然与文化。每个建筑行为(如一系列位于修复建筑低收入公寓的题词)在四处著名的探索领域都被当做是一个表演者。建筑理论的起源(凭借自身的力量成为一种信息来源)涉及到了"耐用性、方便性以及美"。[4]

在这些前提下,要理解现代主义运动及其理性主义的绑定战略是怎样在短时间内内产生巨大的传播力就不再是困难的。显然,相对于在他们的大多数同龄人来说,Ernst May及其同事在他们的时代处理这些问题是更加成功的,他们的做法能够同时积极地与更广范围内建筑要求相接触。新型建造技术、新开发的材料、新发布的土地分配与占有政策以及修改了功利主义与新型讨论的方式的复兴类型学,或被创造,或被修复,或者只是简单地从其他学科中加以借鉴。一旦这些战略变得问题重重,评论家便会"把婴儿连同洗澡水一起倒掉",然后掉进"设想一种备胎运动"的陷阱中,它如同现代主义,应该对一切做出解释。随后的项目逐个地致力于揭露出建筑学中激进的进步或是长久的变革中失败观点的真像。"我们从来不是后现代主义",他们说。

observe even the most complex phenomena in relative isolation, or as part of a uniform scheme. Now we can't. Not after the appearance of semiotic openness and systems theories; not with the discovery of the Internet and the extremely intricate network of global culture. Our evaluations cannot be but relational, these days.

In this vein, a look at present day urban housing projects can be fed with, among other necessary inputs, the cream of the modernist crop. Instead of describing the virtues and flaws of individual works, one could point out at those items in which each new project appears to dispute its precedents' canon, as a part of a larger, progressive tendency. The question remains: have we moved forward, in relation to the protocols of Frankfurt and Brussels, in these forty years? If so, how could we judge our position, if not against others?

Progress becomes tangible once the parts of whatever is being looked at are dissected, and the principles that bind them to other entities are understood. Fortunately, architecture remains a source of elemental knowledge. As such, it explores within a limited number of problems, both natural and cultural. Every architectural action (such as the inscription of a series of low income apartments in a refurbished building, for example) is put forth as a performer in four well known fields of exploration. The origins of architectural theory (a source of information in its own right) speak of "durability, convenience, and beauty."[4]

Under these premises, it is easy to understand how the Modern Movement and its rationalist's hosing strategy achieved enormous diffusion in a short period of time. Ernst May and his colleagues surely tackled the problems of their time more successfully than most of their contemporaries, and in doing so they actively touched on a broad range of architecture's inquiries simultaneously. New building techniques, newly developed materials, unprecedented policies for occupying and distributing the land, reinvigorated typologies, revised utilitarianism and new ways to discuss it all, were either created or recovered or simply borrowed from other disciplines. Once some of these strategies turned problematic, critics threw out the baby with the bathwater, and fell in the trap of assuming a sort of "spare-tire movement" that, just like modernism, should account for everything at once. Individually, the projects that follow contribute to debunk that failed idea of radical progress or permanent revolution in architecture. "We have never been post-modern!" they appear to say.

Operating on the essentials of the rationalist dwelling program, as we can see in these fine examples, contemporary practices introduce additional yet singular elements, in their attempt to make

圣玛利亚住宅遵循一种特殊的建筑语言，这种语言能够将居住行为与一系列精心选择的记忆联系在一起
Santa Maria Housing, following a particular language that is able to tie the act of inhabiting with carefully selected memories

俄罗斯方块：社会住宅和艺术家工作室通过各式各样的抬升，来完成城市区域内多样化和趣味化的需求
Tetris, Social Housing and Artist Studios, fulfilling the need for variety and fun in urban area by making various elevations

　　就像我们在这些优秀的例子中看到的，当代实践引入了额外的非凡元素，作用于具有理性主义的居住项目的基本要素中，从假设的角度来试图使他们的观点比其他相似的建筑理论更健全。一些探索是进步性的，其他的探索则保留着残余的优良传统。但是所有的探索都具有积极的进步观点，并致力于在特殊的环境下，以更好的方式来适应人类生活。

　　探索改进了现代主义的持久性（主要是理性和有效性），例如，Aedes工作室在索菲亚设计的红苹果住宅公寓用砌砖挑战了前人重要的白板理念。建筑师们关于建筑地理位置的抱怨也很有说服力：因为缺少历史层面的意义而使城市环境变得很贫乏；要求抹掉建筑与历史的衔接感的建筑运动也对倡导者提出难以想象的要求。另一方面，阿根斯·伯纳德·布勒设计的"Plantoun的小村庄"也在建筑技术上给予了关注。像过去年代的前辈们一样，这座建筑如同乔治·华盛顿·斯诺于19世纪设计的轻型木构架的当代版本，也希望利用有限的建筑用料来达到绝对的利用效率。对于30所位于巴黎的社会性住宅，KOZ建筑师事务所也选择了木材作为创作灵感的来源。在这个案例中，第二层建筑建在原来的建筑之上，充分地改善了最初的布局，让空间概念和功能体验更丰富。

　　第二组项目在功能主义的背景下，探求了便利性的问题。由Hierve设计的圣玛丽亚宅，假定了其活动性超出简单的功能性。根据他的建筑模式，除了建筑主体外，建筑要允许理念与重要的传统重新联系在一起。建筑形式不仅要遵循功能，也要遵循一种特殊的建筑语言，这种语言能够将居住行为与一系列精心选择的记忆联系在一起。在更实际的层面上，Jaques Moussafir建筑师事务所设计的项目——俄罗斯方块：社会住宅和艺术家工作室，意识到了单调的危险。因为明显的预算限制，即使在最简单的解决办法中，他们的布局策略也在功能性的讨论之中引入了额外的因素：除了舒适性和有效性，建筑也要满足多样性和趣味性。建立在相同的观念上，由Habiter Autrement以及Jean Nouvelle工作室设计的洛尔蒙住宅和LEM+建筑师事务所建造的44所社会住宅单元，显示出一个或两个主要的决定性要素（在

their propositions hypothetically sounder than other similar architectures. Some explorations are progressive, while others rescue the remnants of a tradition that remain valid. All share, however, a positive view of progress and the aim to accommodate human life in the best possible way, under particular circumstances.

The explorations evised modernist durability (its rationality and efficiency, mostly). For example, Aedes Studio's Red Apple Apartments in Sofia uses brick masonry to challenge a crucial aspect of the forefathers' tabula rasa. The complaint raised by the architects regarding their location is eloquent: urban environments are impoverished by the lack of older layers of signification; an unimaginable claim is raised for the masters of the movement that aspired to burn the bridges with history. Agence Bernard Bühler's Small Village of Plantoun, on the other hand, also focuses on building technology. As its predecessors of the past century, it attempts to achieve absolute efficiency in the use of an extremely limited material repertoire, in what could be taken for the contemporary version of George Washington Snow's 19th century balloon frame. In their 30 Social Housing in Paris, KOZ Architectes have chosen wood as a motive, too. In this case, a second layer of building has been grafted upon an older original one, substantially modifying the initial configuration in order to make the spatial perception and the functional experience richer.

A second bundle of projects probe the question of convenience, against the backdrop of functionalism. Santa Maria Housing, a project by Hierve, assumes activities that transcend simple use. According to his model, aside from fostering the body, architecture must also allow the mind to reconnect with important traditions. Form does not follow function only; it also follows a particular language that is able to tie the act of inhabiting with a set of carefully selected memories. At a more pragmatic level, Tetris, Social Housing and Artist Studios, by Jaques Moussafir Architectes, recognizes the perils of monotony. Even in the simplest of solutions, and with evident budgetary constraints, their configurational strategy introduces an additional element into the functional discussion: aside from being comfortable and useful, buildings should also fulfill the need for variety and fun. Based on similar precepts, both Lormont Housing, by Habiter Autrement and Ateliers Jean Nouvel, and 44 Social Housing Units by LEM+ Architectes reveal how the solution of one or two obvious determinants (in this case the quest for individuality and the importance of the views from the site, or the avoidance of noise and other nuisances) can complete projects atop the underlying structure of the modernist program in technical and aesthetic terms.

Icod的12所住宅通过进行简单的类型学实验，来突出项目内在空间的限制，并使其呈现出多样化
12 Houses in Icod, enhancing and multiplying the project's inherent spatial limitations by appealing to simple typological experiments

这个案例中是对个性和场地视觉的重要性的寻求和避免噪音）是怎样在现代主义规划的内在结构之上来完成整个建筑项目的（从技术和审美的层面）。

对于美学的问题也有新的解决方法。毫无疑问，最近对于柏林人的"创新的战前房屋遗产"、维也纳人的"市政房屋住宅"、莫斯科人的"纳康芬公寓楼"等建筑的旧规则提出正式的修改。5468796建筑事物所和Cohlmeyer建筑有限公司设计的"中央村落"是所有项目中非常有趣的一个。虽然保持了以理性主义模式为创造来源的表象，毫无疑问，它还是退后了一步，且面临着一种才思枯竭状态，这种状态起源于通过恢复旧价值和更复杂的群体战略所产生的标准化。DAO建筑事务所在Icod建造的12所房屋，Alfredo Payá Benedito在西班牙阿利坎特省San Vicente del Raspeig设计的32所房屋试图利用简单的类型学实验来提高和增加项目内在的空间极限。错位设计、碎片、局部水磨、重构的确会比传统的隔间更能够突出居住者感觉上的强烈感。

显然，这些项目都不是革命性的。相反，它们逐个地改革或颠覆了现代主义建筑风格的一个或几个基本原则。因此无法在单独层面上找到他们的意义。相反，这些项目将他们的重要性建立在与彼此以及以前的建筑相联系的基础上。每一个都集中于调整和打磨一件或者更多件建筑工具，同时建筑师还提出，建成环境不仅仅是一系列单独的活动，还可以被解读为"人造工艺品"。

麻省理工大学教授斯坦福特·安德森，在1971年发表于《Casabella》杂志[5]的一篇才华横溢的文章上写到："从1930年到1950年，'现代主义运动的主导者'共用的观点是设计过程导致了'设计客体'，也就是说，导致了一个能够根据清晰的、可预见的计划来接受它永久形式的客体。设计师们假设一种理想的模式，然后使它们的形式在这个世界表现出来。这样的'形式赋予者'表现出了一种独裁主义，其经常性的、明显没有充分适应地理位置而设计出的形式，使一切更让人难以忍受。"他还说，"在20世纪60年代，越来越多的社会批评家、建筑师和反对建筑学的设计师们致力于研究建筑师Fuller的项目案例，而规划师和工程师声称城市，甚至更小范围的建筑

There are also new approaches to the question of beauty; formal revisions to the old canon of Berliner Siedlungen, the Viennese Gemeindebauten, and the Muscovite Narkomfin being put forth these days, no doubt. Center Village, by 5468796 Architecture and Cohlmeyer Architecture Limited is certainly a very interesting one, among these reinterpretations. While maintaining a superficial image that undoubtedly feeds on the rationalist model, it takes a step back and confronts the impoverishment that stems from standardization by recovering the value of older and more complex grouping strategies. Also dealing with form, DAO's 12 Houses in Icod and Alfredo Payá Benedito's 32 Housing in San Vicente del Raspeig seek to enhance and multiply their projects' inherent spatial limitations by appealing to simple typological experiments. Dislocation, fragmentation, sectional riffling and reconstitution surely provide sensory intensity to the dwellers of what would otherwise be a conventional set of compartments.

Clearly, none of these projects is revolutionary. Instead, they all reform or subvert one or more of the basic tenets of modernist architecture piecemeal. Their value is thus not to be found at an individual level. On the contrary, these projects base their importance on the way that they relate to each other and to their predecessors. While each appears to focus on tweaking and honing one or more of architecture's tools, together they propose that, more than a series of isolated events, the built environment can be read as an artifact.

"From the 1930s through the 1950s," says MIT professor Stanford Anderson, in a brilliant article published in *Casabella* in 1971[5], *"the 'Masters of the Modern Movement' (...) shared the notion that the design process issues in a 'design object' – that is, in an object that receives its permanent form according to a clear, pre-visioned plan. Designers assumed an idealist position, projecting their forms upon the world. Such 'Form Givers' displayed an authoritarianism that made all the more unendurable by the frequently patent inadequacy of the form to the situation. During the 1960s,"* he continues, *"in increasing numbers and with intensifying fervor, social critics, architects, anti-architectural designers devoted to the example of Fuller, planners and engineers proclaimed that the city – and even smaller, architectural environments – are not design objects fulfilling a stated and persisting program."* And he concludes: *"We should recognize such environments as organizations of form that are the (often unforeseen) result of many human actions, as environments that must sustain a wide range of (often unforeseen) human actions. Such an organization of form, in contrast to an object that is the result of a deliberate design, has been termed an 'artifact'[6]."*

照片提供：© Sir James

柏林人的"创新的战前房屋遗产"位于柏林的Wedding区，由建筑师Bruno Taut于20世纪20年代开发
Berliner Siedlungen in the Berlin district of Wedding, developed in the 1920s by architect Bruno Taut

环境，并不仅仅是设计一些规定的且长久的项目的客体。"他总结到："我们应该把环境当成组织形式，这种组织形式（通常难以预见的）是人类行为的结果，因为环境必须承受大量的（难以预见的）人类行为。这样的组织形式对比那些刻意设计的客体，被定义为'人造工艺品'⁶。"

这些问题，而不是其他建筑风格上的问题，才是矛盾的真正要素。这是我们时代的标志。我们依旧被世纪的准则束缚，可是我们仍旧认为自己是进步的。个体建筑师，即力量强大且控制着自己的设计的设计师，成为现代建筑规划的基础，依旧获得了巨大的成功。文艺复兴的倡导者、古典建筑最高奖的获得者、革命性的理想主义者和现代主义的城市设计师们和那些接下来和我们分享他们作品的同事们仅仅有些许不同。

但是，他们的"人造工艺品"超出了项目本身，为进步开启了前所未有的大门。成群的、交织在一起的以及互相联系的当代事件不应该被某个女人或男人的静态图片所展示，也不该被他们设计的一系列作品来展示，而应该由依赖于建筑地点和建筑原则的复杂网络展示出来，这些地点和原则允许我们塑造出建筑环境，同时促使对建筑实践和可能的作品进行新的阐释。

斯坦福·安德森在另一篇文章中说，"尽管存在着一些特别的批判，我想我们应该认同本世纪公共住宅取得了进步……我们从提议、检测、再造，或者摒弃简单和不足的假设，比如'生存最低限'中学得经验。在开发大规模的、建立在这些假设上的房屋项目中运用非凡的批判意识，会给我们提供更有用的信息"。⁷

所以，我们可以看到，这些建筑不止是无名居住者的城市房屋，它们还展示了不断进步的、健康的有机体系的生成原则。面对如此复杂的创造物，Aymonino的批判工具显然有点不合时宜了。

These, not other stylistic matters, are the true elements of contradiction. This is the sign of our times. We remain bound by centennial canons, and yet assume ourselves as progressive. The figure of the individual architect, the ever-powerful designer who remains in control of his projects, underlies the modern architectural program and still operates with surprising success. Renaissance masters, Beaux Arts Grand Prix laureates, revolutionary utopians and Modernist urban designers differ but slightly from the colleagues who proudly share their work with us in the following pages.

Beyond the projects themselves, however, it is their artifactualness which opens an unprecedented door to progress. Clustered, articulated and interrelated, the events of our time should be represented not by the static picture of a man or a woman and the series of objects they produce, but rather by the complex map of a network fed by positions and principles which allow us to configure the built environment while promoting a new interpretation of the architectural practice and its possible products.

"Despite only rather ad hoc criticism," says Stanford Anderson, in another article, *"I think we would agree that public housing has improved during the century. (...) We learn from the proposal, testing and reformulation or rejection of simple and apparently inadequate hypotheses such as the Existenzminimum. The employment of a greater critical awareness during the development of large-scale housing projects which are based on such hypotheses might have provided us with much usable information."*⁷

So seen, more than urban houses for anonymous dwellers, what follows is a show of the generative principle of a healthy organism in constant evolution. In dealing with such a complex creature, Aymonino's critical apparatus does seem a bit obsolete.

Jorge Alberto Mejía Hernández

1. Adolf Loos, "Potemkin City(July 1898)", in *Spoken into the Void: Collected Essays 1897–1900*, Cambridge(Mass.): The MIT Press, 1982, p.95
2. Carlo Aymonino, *L'Abitazione Razionale: Atti dei Congressi CIAM 1929-1930*, Padova: Marsilio, 1971
3. "Modern architecture died in St Louis, Missouri on July 15, 1972, at 3.32pm (or thereabouts), when the infamous Pruitt Igoe scheme, or rather several of its slab blocks, were given the final coup de grace by dynamite." Charles Jencks, *The New Paradigm in Architecture: The Language of Post-Modernism*, New Haven and London: Yale University Press, 2002 (7th edition of The Language of Post-Modern Architecture)
4. Vitruvius, *The Ten Books on Architecture*, New York: Dover Publications Inc., 1960, p.17
5. Stanford Anderson, "L'Ambiente Come Artefatto: Considerazione Metodologiche / Environment as Artifact: Methodological Considerations", Casabella, No.359-360, year XXXV (1971), pp.71~73
6. Footnote 4 on the original: "F. A. Hayek, "The Results of Human Action but not of Human Design", in his *Studies in Philosophy, Politics and Economics*(pp.96~105, London: Routledge and Kegan Paul, 1967)", in Ibid, p.77
7. Stanford Anderson, "Architecture and Tradition that Isn't 'Trad, Dad'," in Marcus Whiffen(ed.), *The History, Theory and Criticism of Architecture*, Cambridge(Mass.): The MIT Press, 1965, pp.87~88

洛尔蒙住宅
Habiter Autrement + Ateliers Jean Nouvel

2012年，Habiter Autrement和Jean Nouvel工作室因为波尔多的洛尔蒙住宅而被授予Agola建筑奖。以MVRDV建筑事务所的Winy Maas为首的评审团在阿基坦桥下的场地上领略到了项目的美和复杂性。

在洛尔蒙住宅项目中，吉伦特河与阿基坦桥的超豪华视野成为场地的关键性特点。为了优化这些视野，同时使邻居间的可视性最小化，每间公寓（共26间）都被赋予了独特的外形。这座建筑的整体布局是一个锯齿形状。

场地的朝向赋予了建筑两条长长的侧边，面向东侧和西侧，各具特色。沿着多面的西立面，不透明的嵌板与全高的窗户交替呈现，以提供河流的全景视野。开放的窗户将起居空间改造为一处室外露台。剖面设计强化了60m高的桥的存在，使其看起来非常壮观。两层公寓叠加起来，地面层以上设有复式公寓和三层公寓。每一层横向都具有流动性，使部分屋顶暴露出来，允许全高的窗户延伸至屋顶。这一立面上的垂直玻璃条到横向屋顶照明的转变突出了低角照射（使用电影术语来说）的效果，使其向上面对着桥塔。

东侧，建筑通过一个带有小型洞口的金属嵌板立面，来远离嘈杂的街道。洞口的规格与位置是由公寓内的需求所决定的。横向通道覆以反射材料，使其阳光充足，创造一幅模糊的、梦幻般的河流形象，人们在街道一侧便可欣赏这一处场景。这条通道还加以拓宽，鼓励各种欣赏河流的方式。

Lormont Housing

In 2012, Habiter Autrement and Ateliers Jean Nouvel were awarded the Agora Architecture Prize for the Lormont Housing in Bordeaux. The jury, headed by Winy Maas of MVRDV, recognized the beauty of the project as well as the complexity of building on a site under the Aquitane Bridge.

In the Lormont Housing project, extraordinary views of the Gironde River and the Aquitaine Bridge are the key features of the site. In order to optimize these views, while minimizing visibil-

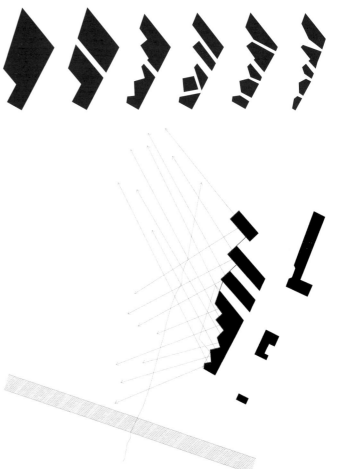

Aquitaine River

riverside walk

river view — flowering prairie — river view — solar panels

glazed roof — passage — passage — passage — passage — private terrace

planted parking roof

Jean Bonnin Street

Chaigneau embankment

0 20 50m

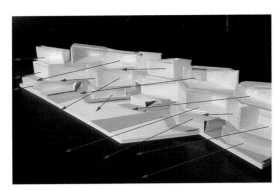 

A-A' 剖面图 section A-A'

B-B'剖面图 section B-B'

项目名称：Lormont Housing
地点：Lormont, France
建筑师：Habiter Autrement_Mia Hägg, Sandrine Forais,
Ateliers Jean Nouvel_Jean Nouvel
项目经理：Aude Lerpinière
项目团队：Aimée Lau, Lina Lagerström, Félix Medina Mathilde Jauvin, Louis Mangin, Nicolas Métro, Marcel Züger
竞赛团队：Pawel Krzeminski, Lina Lagerström, Kasia Jackowska, Noélie Sénéclauze
建筑顾问：Jean-Jacques Raynaud
结构工程师：Khephren Ingénierie
景观：Ewen Le Rouic, Irène Djao Rakitine, Céline Aubernias
甲方：Foncière Logement 开发商：ING Real Estate Development
用地面积：3,514m² 总建筑面积：2,927m² (surface habitable) 有效楼层面积：5,189m² (surface SHOB)
规划：26 apartments
竣工时间：2011
摄影师：©Philippe Ruault (courtesy of the architect)

二层 second floor

ity between neighbors, each of the 26 apartments was given a unique shape; the result of this arrangement is the building's overall saw-shaped figure.

The site orientation gives the two long sides of the building, which face to the east and the west, with very different characteristics. Along the faceted west facade, opaque panels alternate with full-height windows that offer panoramic views of the river, and which open to transform the living space into a terrace. The sectional design works to enhance the spectacular presence of the 60 m high bridge. Two levels of apartments are stacked, with duplexes and triplexes above the ground floor. Each level is shifted laterally in order to expose part of the roof, allowing the full-height windows to extend into the ceiling. This transition from the vertical strips of glass on the facade to the horizontally-orientated roof lights emphasizes the effect of the "low-angle shot", to use a cinematographic term, up towards the bridge towers.

On the east side, the building is shielded from a noisy street by a metal-clad facade with relatively small openings, the size and position of which are determined by the needs of the apartments within. Transverse passages, clad in reflective material, amplify available light and create a blurry, dreamlike image of the river that is visible on the street side; the passages also widen to invite the approaches towards the river.

三层公寓 triplex apartment

复式公寓 duplex apartment

四层 fourth floor

三层 third floor

三层 third floor

二层 second floor

二层 second floor

复式公寓 duplex apartment

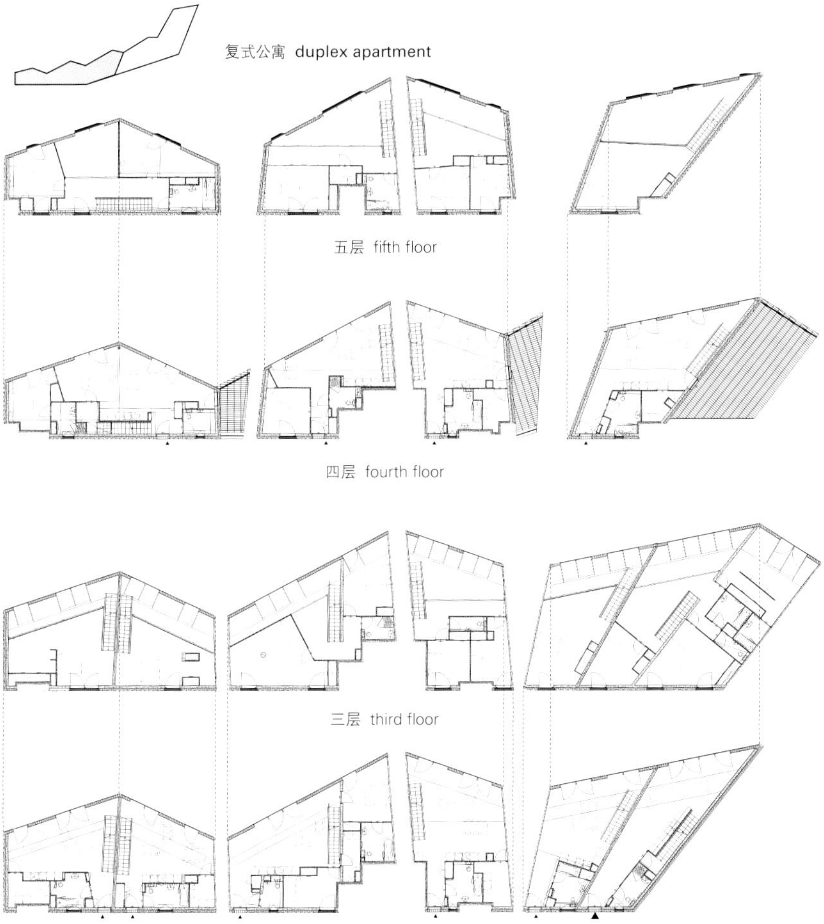

五层 fifth floor

四层 fourth floor

三层公寓 triplex apartment

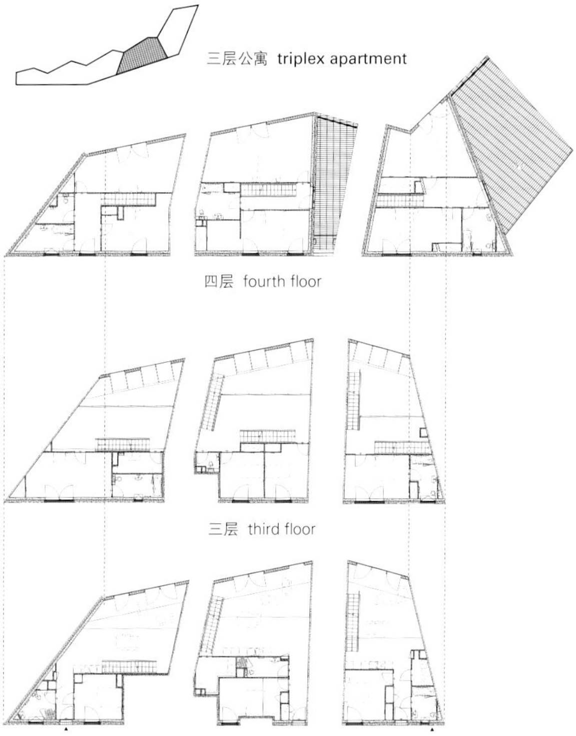

四层 fourth floor

三层 third floor

三层 third floor

二层 second floor

二层 second floor

三层公寓 single apartment

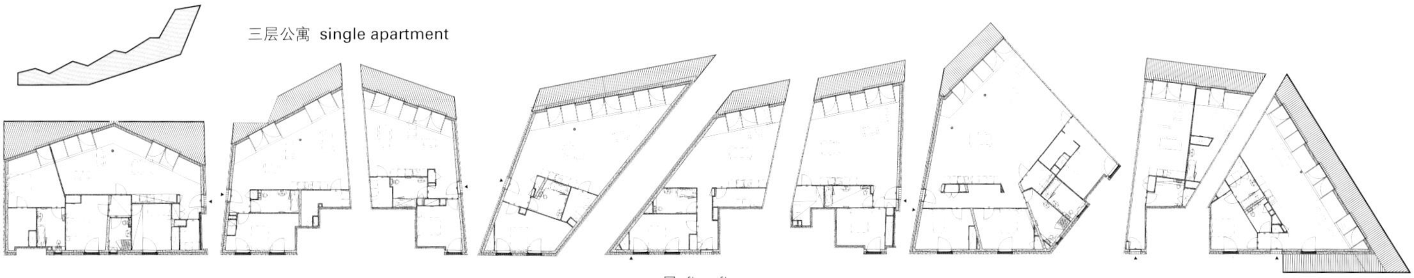

一层 first floor

Icod的12所住宅

DAO

早前建筑师试图设计一个布局清晰且充满活力的大楼,人们在这里能和谐相处。一个特定的起源能赋予建筑更加长久且持续的内容。

Icod是一座拥有12个住宅的混合体,位于公寓大厦与带有花园的独立住宅之间。但是这只是一个结果。这座建筑具有两个罕见的特点。一是其地理定位,另外一个特点则是它被当地的私人业主所主导。

特内里费岛是一个多山的岛屿,人口密度约为500人/km^2,较高的土地占有率以及较低的人口密度使该区域的可持续性过渡饱和,呈现出绝对的优势。来自于建筑师与开发商之间协商的机遇产生了一些绝对合适的战略。

对于建筑师来说,这个岛屿是一处自主且自给自足的土地,是一座由住宅构成的城市,领土转变为城市风格,这是它的风格,建筑师把它叫做城市领土化。建筑师利用材料,且与本地供应商合作,在这个框架中生成一个节点,但是却在新范例的内部产生了一个密度更大且效率更高的单元。

这一切导致了街区内的私营推广来强化当地习惯的过程。

建筑师旨在加强当地居民之间的内在关系。从能效方面来说,一方面他们尽可能地寻求高密度,以及最大数量的住宅;另一方面,场地的巨大深度(陡峭的斜坡)则暗含了庭院位于其内部。这两个原因使建筑师尝试一个单一的战略,这足以证明他们的野心:过渡开发这些庭院,使它们的使用率密集化。

一个带有庭院的住宅抬高了其中的一处,使小径可以穿过庭院,位于廊道之下。廊道之后会成为一处楼梯。这个方案与其相邻的住宅一起产生了一种强烈的共生关系,以此来进行自身调节,以及促进彼此的关系。

这座建筑是由6个重复设置的模块构成的,模块实现了高能量的利用率,以形成一处加以控制的、阴凉的且受保护的公共空间,它是整座建筑的肺部和心脏。这座建筑是两座由通道连接的建筑构成的。每座住宅都平均地分配到这两栋建筑之内,此外,每一半的住宅还处在不同的楼层。这个规划以一种自然的方式将社区生活引领至屋顶层,将这处别具风格的空间作为室外游乐空间(花园),从而促进整体质量的提升。

住宅是有着两三个小规格房间的房屋,但是空间却十分宽敞。所有住宅的朝向、视野、高度、通道、通风以及照明条件都是一致的。因此,建筑师根据项目暗含的无限循环的假定条件,来实施这个方案。

12 Houses in Icod

Early on the architects try to get a clear design and an alive building where inhabitants can work out relationships. A specific genetics provides the building the longer lasting content.

Icod is a hybrid of 12 houses between a block of apartments and detached houses with garden. But this is only a consequence. What explain it are 2 singularities. The one is territorial geoloca-

tion, as for another, it is promoted by a local private initiative. Tenerife is a hilly island with a population of about 500 inhabitants per km², but with a very high rate of land occupation with low density being ripe to oversaturate and to overwhelm its sustainability. The opportunities come up of the negotiation among the architects and the developer to get the appropriate strategies.

For the architects, the island is a field of autonomous and self-regenerating cells as a city made out of houses; territoriality turned urbanity; that is its scale, they call it terrurbanity. And they create a node in this frame working with materials and local suppliers, but generating a denser and more efficient cell with a new paradigm within it.

All this led to a process of empowerment of the local habits in private promotion for the neighborhood.

The architects aim also to get an inner empowerment for the relationships of the inhabits. In terms of efficiency they look for the highest density possible, the maximum number of dwellings, on one hand. And on the other hand the great depth of the plot (steeply sloping) presaged courtyards inside it. Both reasons led them to implement a single strategy which proved enough for all the ambitions: overexploit these courtyards, densify their use. A courtyard house that raises one piece of it allows passage through the courtyard under the corridor that then became stairs. This scheme along with its neighbor, produces a symbiosis so strong that it autoregulates and promotes relationships.

This module repeated 6 times generates the building and achieves high – energy use efficiency, in order to obtain a controlled, shaded and protected public space that serves as lung and heart. The whole building is 2 buildings connected by jetways. Then one half of every house is in one building and one half in the other building, besides they are in different floors. This scheme leads the community life to the roofing in a natural way, providing this space of idiosyncrasy as exterior play area (the garden) and qualifying the whole promotion.

They are homes with 2 and 3 rooms of small dimensions but with an enormous spatiality. All houses have identical conditions of orientation, view, height, access, ventilation and lighting. Consciously the architects have acted since metabolist postulations were implicit in the project. DAO

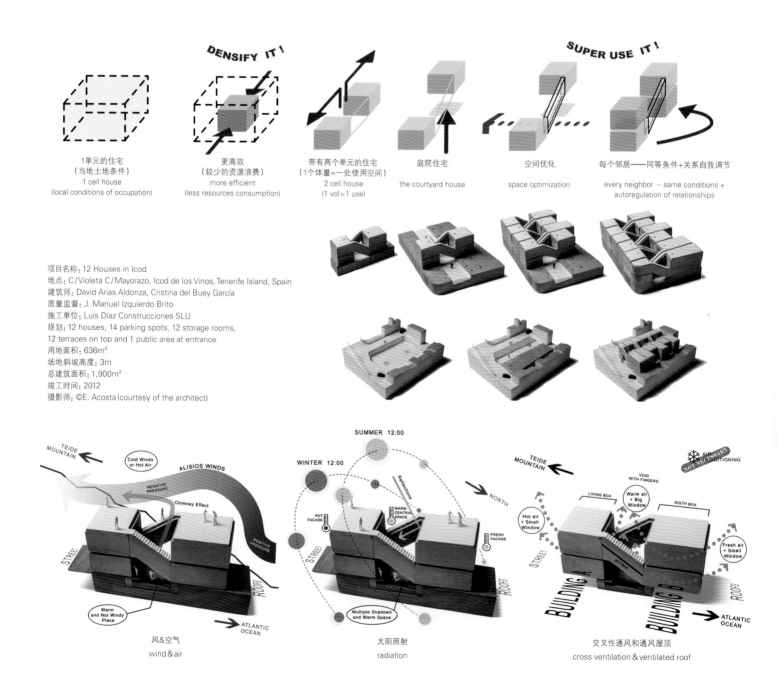

项目名称: 12 Houses in Icod
地点: C/Violeta C/Mayorazo, Icod de los Vinos, Tenerife Island, Spain
建筑师: David Arias Aldonza, Cristina del Buey García
质量监督: J. Manuel Izquierdo Brito
施工单位: Luís Díaz Construcciones SLU
规划: 12 houses, 14 parking spots, 12 storage rooms, 12 terraces on top and 1 public area at entrance
用地面积: 636m²
场地斜坡高度: 3m
总建筑面积: 1,900m²
竣工时间: 2012
摄影师: ©E. Acosta (courtesy of the architect)

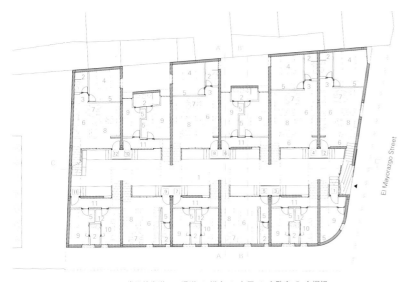

1 常见的街道——通道 2 浴室 3 大厅 4 主卧室 5 衣帽间
6 厨房 7 餐厅 8 起居室 9 卧室 10 工作室 11 交通流线区域 12 架子区

1. common street – accesses 2. bathroom 3. hall 4. main bedroom 5. wardrobe
6. kitchen 7. dinning room 8. living room 9. bedroom 10. studio 11. circulation area 12. shelf

一层 first floor

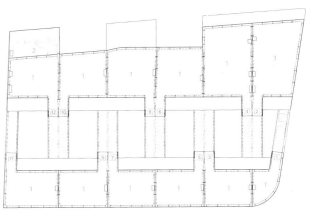

1 露台游玩区 2 空中区

1. terrace – play area 2. aerials area

三层 third floor

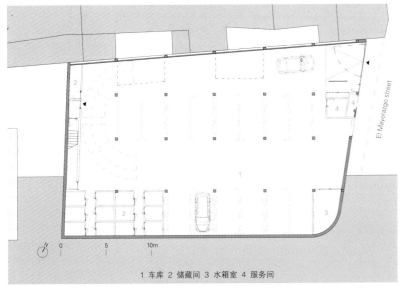

1 车库 2 储藏间 3 水箱室 4 服务间

1. garage 2. storage 3. water tank room 4. service room

地下一层 first floor below ground

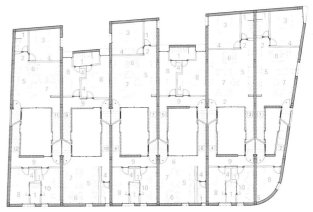

1 浴室 2 大厅 3 主卧室 4 衣帽间 5 厨房 6 餐厅
7 起居室 8 卧室 9 交通流线区域 10 工作室 11 架子区

1. bathroom 2. hall 3. main bedroom 4. wardrobe 5. kitchen 6. dinning room
7. living room 8. bedroom 9. circulation area 10. studio 11. shelf

二层 second floor

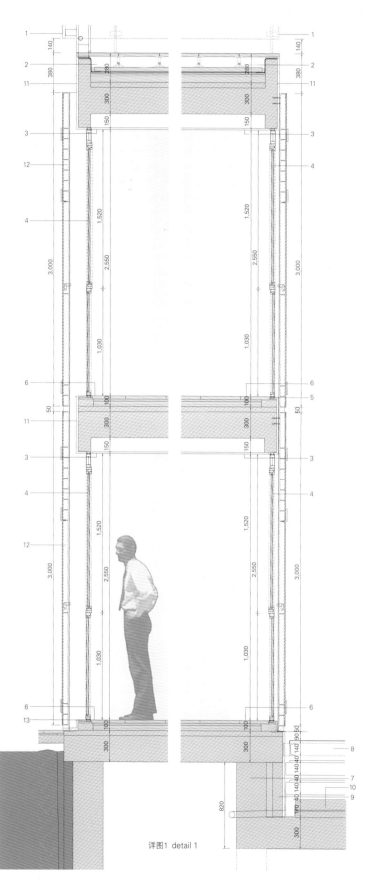

详图1 detail 1

1. railing
   - enclosure: stainless steel cable mesh x-tend by carl stahl
   - structure: horz. galvanized steel pipe, d=40mm, d=20mm vert. double galvanized steel pair t=10mm, w=40mm all with 2 layers of anticorrosive painting+2 layer of black exterior steel painting
2. ventilated flooded roof
   - interior sandwich: one direction slab with concrete block as lightened t=25+5 cm/acoustic sheet impact sound insulation t=6mm/foamed concrete as slooping 0%, t=10cm+ trowelled mortar layer t=2cm/EPDM rubber membrane as waterproofing, t=1,8mm/geotextile sheet, t=2mm
   - exterior sheet: filtron slab tiling by intemper with insulation of extruded polystyrene. the tilling lay over special foamed concrete blocks as plots
3. suspended plaster ceiling with 12cm for facilities
4. anodized aluminium metal work (1st class) with double glazing 6+6+6mm
5. aluminium weatherboard
6. ground floor and 1st floor slabs: ceramic tiling over 3 cm of micromortar/acoustic sheet impact sound insulation t=6mm/one direction slab with concrete block as lightened t=25+5cm
7. edge beam of reinforced concrete
8. formworked reinforced concrete steps at entrances to the houses
9. enclosure on facades: wall of concrete blocks 9x25x50cm/to interior. rough render t=1,5mm+finishing rendering t=1,5mm with 2 layers of white plastic paint for interior/to exterior. white monolayer mortar coating (fine finishing) t= 30mm with 3 layers of white paint for facades
10. public inner street slab: continuous concrete pavement. polished finishing t=5 cm/foamed concrete as slooping 2% t=10cm/geotextile sheet, t=2 mm/EPDM rubber membrane as waterproofing, t=1,8mm/one direction slab with concrete block as lightened t=25+5cm/lineal caves gutter made in situ. oping 4% coated with EPDM membrane
11. monolayer coating smooth finishing over reinforced concrete breastplate
12. aluminium shutter for shading and privacy made of oval tubes to get minimum maintenance
13. exterior sidewalk of hydraulic tiling 25x25cm over mortar and sand over concrete wall plate

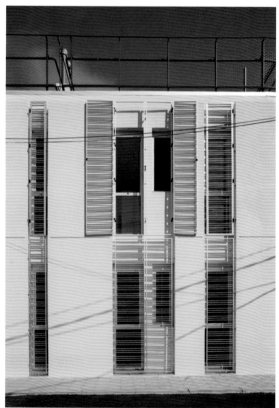

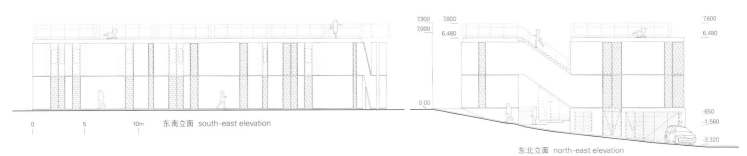

东南立面 south-east elevation

东北立面 north-east elevation

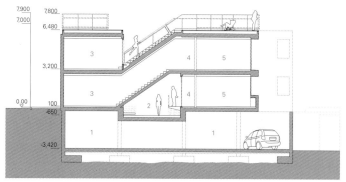

A-A' 剖面图 section A-A'

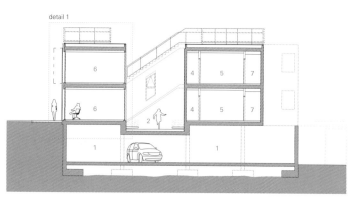

B-B' 剖面图 section B-B'

1 车库
2 常见的街道——通道
3 起居室
4 交通流线区域
5 卧室
6 厨房
7 浴室

1. garage
2. common street – accesses
3. living room
4. circulation area
5. bedroom
6. kitchen
7. bathroom

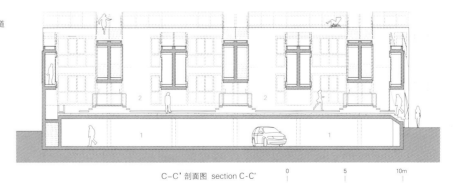

C-C' 剖面图 section C-C'

San Vicente Del Raspeig的32所住宅

Alfredo Payá Benedito

☐ facing conservatory
☐ lattice of wooden slats
☐ raw concrete
☐ rough plaster with arid
☐ local enclosure

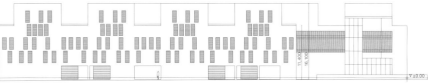

西北立面 north-west elevation

西南立面 south-west elevation

这处城市规划是由交流空间连接的两座平行建筑体块构成的。市议会对此表示接受，且连同其建筑师重新思考场地建筑体块的布局。

他们所做的第一个决策便是移走体块，这种方式产生了新型的渗透性，且建筑所处区域的影响力也有所扩大。建筑体块之间的空间，即一座广场和一条覆顶的街道，成为交通流线中的一部分。

第二个决策便是在一层设置商店，建筑师利用开放式的平面布局来平衡面向大道、花园以及覆顶广场一侧的体块。

整座住宅建为两层，这样，入口便实现了最小化。入口处种有植被，是日间使用空间和夜间使用空间来回交替的场所。住宅的每个房间都实现了自然照明和对流通风。

叠加的战略允许一半以上的住宅（16所）位于顶层，这些住宅被称为露台住宅，其余的住宅（16所）则沿着覆顶的街道设置。街道作为住宅空间的延伸。通过这种方式，每座住宅都与室外空间保持联系。

由普通变得非凡

住宅另一侧的空间并不是毫无生机、毫无活力的，而是时刻地保持某种活力和变化。通过这种方式，普通变得非凡，每天的生活都在传达和创造与外界、与环境、与城市的联系。

共享空间

这项研究利用住宅建筑的建造优势，来创造能够在街道和住宅之间创建一种亲密关系的集体空间。空间使社交生活重新充满了活力，恢复了邻里之间以及建筑与城市之间的交流。它实现了共享空间的理念，简而言之，实现了一个创造公共空间的机会。

建筑师规划了一个足够大的环境视野，在这里，作为重要元素的地点、周围环境以及天气，成为连接地中海文化的生活方式。露台、覆顶街道以及门廊，都成为了情感空间。

32 Housings in San Vicente del Raspeig

The urban proposal was two parallel blocks connected by communication spaces. The receptivity of the city council and its architects allowed to rethink the layout of the blocks on the site.
The first decision was to move the blocks, in this way new permeabilities appeared and the area of building influence increased. The space between the blocks, a square and the covered streets, assumes the role of circulations.
The second decision was to choose to occupy the ground floor with shops, through an open-plan that balances the blocks facing to the avenue, to the garden, and to the covered plaza.
The whole housings have two stories, in this way the accesses are minimized. In these accesses plants are always daytime spaces and up and down alternately the nightime spaces. Every room of housing has natural light and cross ventilation.

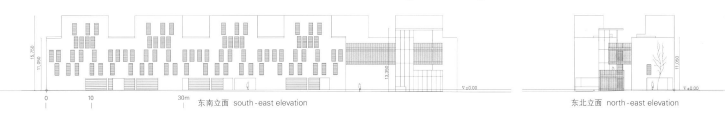

东南立面 south-east elevation　　东北立面 north-east elevation

五层 fifth floor

四层 fourth floor

三层 third floor

二层 second floor

一层 first floor 0 10 30m

2单元_三层 unit 2_third floor

2单元_二层 unit 2_second floor

4单元_五层 unit 4_fifth floor

4单元_四层 unit 4_fourth floor

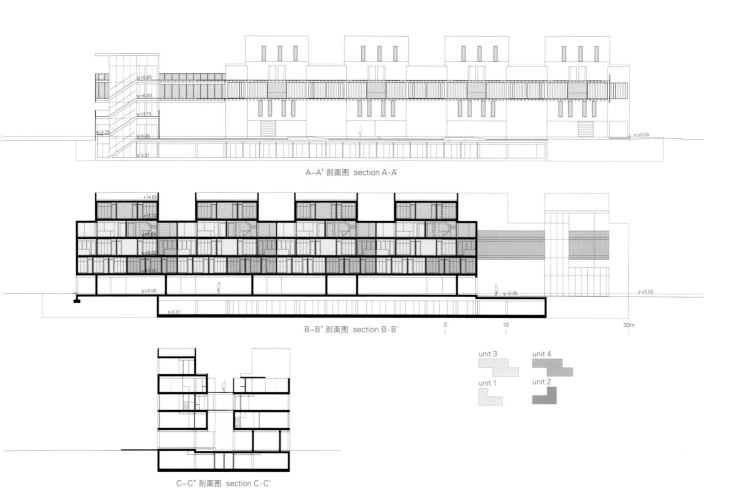

A-A' 剖面图 section A-A'

B-B' 剖面图 section B-B'

C-C' 剖面图 section C-C'

unit 1　unit 2　unit 3　unit 4

项目名称：VPO en San Vicente
地点：San Vicente del Raspeig, Alicante, Spain
建筑师：Alfredo Payá Benedito
合作商：Raquel Del Bello, Gema Vicente, Arturo Calero Hombre, Sonia Miralles Mud, Vicente Pascual Fuentes, Marcos Gallud García, Natalia Velasco Velázquez, Beatriz Vera Payá
机械工程师：Juan Jesús Gutiérrez Sánchez
土地所有权：Instituto Valenciano de la Vivienda S.A.
施工单位：Cántera Vértice S.A.
有效楼层面积：6,063.08m²
造价：EUR 2,761,103.05
设计时间：2003
施工时间：2005.10
竣工时间：2012.1
摄影师：©David Frutos (courtesy of the architect)

1. no transit roof: rave 20/25 3cm geotextile waterproof sheet fiberglass/regularized layer concrete mortar 2cm/formation of slopes concrete 5cm
2. outer protection: galvanized steel cross board tube 40x40mm/ fixed to steel structure according to detail/cooper treated fine width 12cm
3. reinforced concrete stairway
4. elevator: galvanized steel sheet
5. outer paver: reinforced concrete slab 10cm beams structure reinforced concrete 15x25cm
6. grid paver: galvanized steel grid rectangular void 4cm/frame L 50.50.5
7. outer protection: galvanized steel tube 60x40cm screwed to slab according detail crossbar galvanized steel tube 40x40xm cooper treatement pine plank width 12cm separation between planks 6cm
8. outer paver: reinforced concrete slabs 5cm/ protection mortar/waterproof sheet/pendents of slopes concrete/reinforced concrete waffle/ suspended false celling
9. gride paver: galvanized steel grid rectangular void 4cm, frame L 50.50.5 neoprene
10. plaster false celling/painted in white
11. underground paver: polish concrete 10cm waterproof sheet/concrete pendents of slope reinforced concrete slab/graves
12. grid faver: galvanized grid rectangular void thickness 4cm/ frame 50.50.5
13. outer paver: reinforced concrete slabs 5cm reinforced concrete beams structure 15x30cm
14. parapet: steel plate 5cm
15. garden roof: grave/mortar/impermeabilized sheet concrete reinforced concrete waffle slab plaster false celling

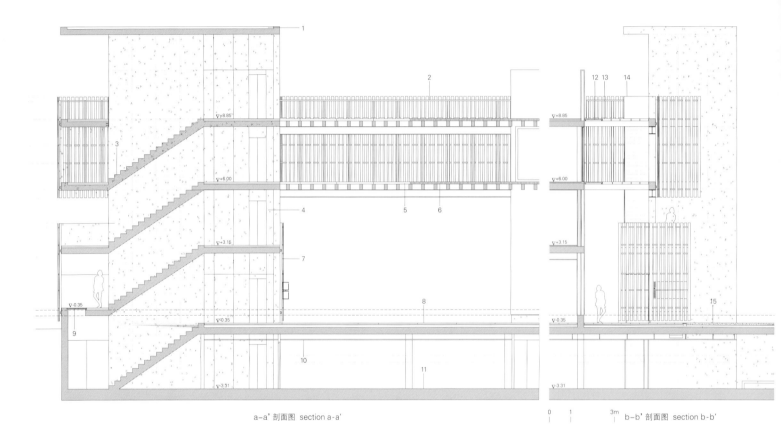

a-a' 剖面图 section a-a'   b-b' 剖面图 section b-b'

详图1 detail 1

1. protección exterior: travesaños de tubo de acero galzinizado 40x40mm anclados a la estructura metália según detalle separación entre tablas aprox. 6cm
2. pavimento exterior: losas de hormigón armado de 10cm losa de H.A de 25cm
3. tabique ladrillo perforad del 12 enfoscado a 2 caras, rugoso al exterior, color a elegir por la D.F.
4. pavimento exterior: terrazo grano fino 40x40cm mortero de cemento capa de arena 2cm
5. frente de puertas: acabado de chapa de acero galvanizado lacado en color a elegir por la D.F
6. frente de puerta de ascensor: panelado de chapa de acero galvanizado lacado en color a elegir por la D.A
7. pavimento exterior: losas de hormigón armado de 10cm estructura de H.A de 15x25cm
8. pavimento rejilla: rejilla de acero galvanizado de hueco rectangular espesor 4cm, marco L 50.50.5, sobre tubo 50.50.0 atornillado a forjado
9. escalera con losa y peldañeado de H.A.
10. peto escalera: pletina de acero de 1.5cm soldada a borde de forjado

Stacking strategy allows half of the housings(16) on the top, these are calling housing-patio, and the rest(16) are situated along the covered street. This street works as an extension space of the housing. In this way every house has outdoor space associated.

The ordinary becomes extraordinary

The space remaining on the other side of the house is not inert and lifeless but living and changing. In this way the ordinary becomes extraordinary, everyday life conveys and creates links with the outside, with context, with the city.

Shared Spaces

The research was to take advantage about the construction of a dwelling building in order to create collective spaces capable of establishing a close relationship between the street and the house. Spaces make revitalizing of social life, and the exchange between neighbors and the rest of the city. It appeared the idea of the shared space and, in short, an opportunity to create public space.

The architects propose an amplified view of the context where the location, surroundings, and climate are important, as the lifestyles linked to Mediterranean culture. Patios, covered streets, porches, are emotional spaces. Alfredo Payá Benedito

俄罗斯方块：社会住宅和艺术家工作室
Jaques Moussafir Architectes

这个社会住宅项目的委托者是SIEMP，是一个公营企业，这个企业因材施教的方法极好地响应了修正过的城市复兴战略。与20世纪80—90年代的移除整座体块的战略截然相反，今天的巴黎更倾向于尽可能地保留应有的建筑。

Jaques Moussafir建筑师事务所的俄罗斯方块住宅是巴黎北部一处肮脏街区的一个大型复兴工程的一部分：犯罪和无法接受的居住条件是需要战胜的挑战。城市、当地的协会与业主（包括新住宅与翻修的廉价租房，以及艺术家和音乐家的工作室的业主）亲密合作，来指定该项目的平面布局。

这个项目由城市规划师Patrick Céleste领导，由七个建筑团队共同完成。Jaques Moussafir建筑师事务所负责的三处场地位于两条狭窄的街道上，这两条街道被一处低层住宅区隔开。住宅区中心的花园使该项目的两个部分产生了视觉互动。这一微型城市部分被20世纪60年代的大量建筑所围绕，但是却仍然保留了其郊区特有的精髓和规模。

项目是在尊重邻里街区的规模以及密度的意愿的基础上成形的，同时还为未来潜在的住户尽可能多地提供空间和光线。本项目与其临近的建筑呈线形排列，两座rue du Nord建筑之间甚至还夹着一座家庭住宅。融入环境的意图被室外使用的建筑手法得以强化：石灰泥与周围立面的纹理相呼应，大部分的窗户洞口规格都采用传统的比例。百叶窗被覆以同样灰泥，使窗户处于关闭的状态时，几乎从立面中消失不见。唯一的人工元素便是起居室的扩建：由绿柄桑木建成的大型框架式突出部分。

按照标准建立的社会住宅严重地限制公寓的大小。俄罗斯方块住宅利用一系列的设计方案来使资源利用最大化。场地的紧凑性使每层只建造一座公寓成为可能。一层的公寓适合行动较少的住户，条状的私人微型花园成为一处缓冲区以及一层租户的私人屏障。利用建筑前侧和后侧立面的南北朝向，起居室被赋予了双重性。

起居室的天花板高度为3.40m，而卧室的高度为2.60m，这种不同更加突出了公园的日间区域。这一结果是通过交错排列的、俄罗斯方块式的布局来完成的，二层的卧室置于一层的起居室之上，来为每个单元创造个性化的布局。全高的、木框架式的弓形窗户成为每个公寓独特的标志，同时使起居室面向街道开放。工作室和公寓的设计完全相同，因此模糊了当代住宅与工作空间的区别。

在俄罗斯方块住宅内，平庸和特殊性共存。这个项目承认普通事物的价值，致力于展现城市肌理的连续性，同时通过添加独特的元素来对其进行突出。此外，基本材料的使用也使预算发生了变化，以提高居住空间的质量。

Tetris, Social Housing and Artist Studios

The commission for this social housing project came from SIEMP, a public enterprise whose case-by-case approach responds to the city's revised urban renewal strategy. Contrary to the 1980s-1990s policy of razing entire blocks, today Paris prefers to maintain what can be maintained.

Jaques Moussafir Architectes' Tetris houses were part of a larger regeneration program for a squalid neighborhood in Northern Paris: delinquency and unacceptable living conditions were the challenges to overcome. The action plan was developed in close cooperation between the city, local associations and the landlord included new and refurbished low-rent housing, as well as studios for artists and musicians.

Led by urban planner Patrick Céleste, the project was shared between seven architectural teams. Three plots entrusted to Jaques Moussafir Architectes were sited on two narrow streets separated by a low-rise housing block; a garden in its middle enabled a visual interaction between the two parts of the project. Surrounded

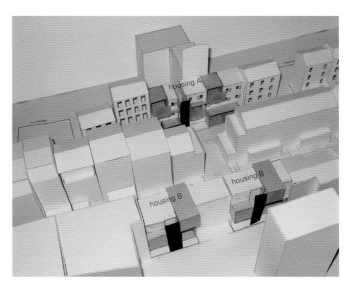

by massive 1960s buildings, this tiny fragment of the city retained its faubourien spirit and scale.

The project was shaped by the desire to respect the neighborhood's scale and density while maximizing space and daylight available to future residents. The houses are perfectly aligned with their neighbors; the two rue du Nord buildings even have a one-family home neatly sandwiched between them. The intention to blend with the context is emphasized by exterior treatment: lime stucco echoes the texture of surrounding facades; most window openings have traditional proportions. The shutters are covered with the same kind of stucco so that closed windows almost disappear on the facade. The only articulated element is the extension of living rooms: large frame-like juts constructed from iroko wood.

Standards established for social housing severely restrict the size of apartments. Tetris uses a series of design solutions to make the most of the available resources. The compactness of the plots made it possible to build no more than one apartment per floor. Ground floor apartments are suited for residents with reduced mobility; a strip of private micro-gardens serves as a buffer and privacy screen for ground floor tenants. Taking advantage of the front and rear facades' North-South orientation, living rooms are given a double aspect.

The difference in ceiling heights – 3.40m for the living room and 2.60 for the bedroom – amplifies the apartment's day zones. This was achieved through staggered, Tetris-like arrangement where the second-floor bedroom is placed upon the first-floor living room, and so forth, creating individualized layouts for each unit. Full-height, wood-framed bow windows serve as distinguishing marks for each apartment, while opening the living room towards the street. Ateliers and apartments are designed identically, acknowledging the blurred boundaries between contemporary home and workspace.

The banal and the exceptional coexist in Tetris houses. Recognizing the value of ordinary things, the project contributes to the continuity of urban fabric, and at the same time enhances it by adding unique features. Besides, the use of basic materials allowed redirecting part of the budget for improved quality of living spaces.

项目名称：Tetris, Social Housing and Artist Studios
地点：13/15 rue du Nord, 19 rue du Nord and 14-18 rue Emile Chaine, 75018 Paris
建筑师：Jaques Moussafir Architectes
设计团队：Jacques Moussafir, Alexis Duquennoy
工程师：SIBAT 承包商：SRC
甲方：SIEMP(Société Immobilière à Economie Mixte de la Ville de Paris)
规划：9 apartments, 3 artist studios
用地面积：449m² (132m² + 106m² + 211m²)
总建筑面积：293m² (79m² + 64m² + 150m²)
有效楼层面积：794m² (218m² + 189m² + 387m²)
材料：concrete masonry, plaster, lime stucco, wood
造价：EUR 1,460,000
设计时间：2004.8~2006.7 施工时间：2008.11~2010.6 竣工时间：2010
摄影师：
©Luc Boegly (courtesy of the architect) - p.57, p.58, p.61, p.62, p.63
©Jean-Claude Pattacini (courtesy of the architect) - p.60

南立面_住宅A south elevation_housing A

housing A, Rue Emile Chaine

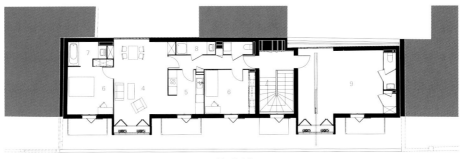

三层　third floor

二层　second floor

一层　first floor

1 大厅
2 垃圾室
3 自行车棚
4 起居室
5 厨房
6 卧室
7 浴室
8 洗衣房
9 艺术家工作室

1. lobby
2. garbage room
3. bike shed
4. living room
5. kitchen
6. bedroom
7. bathroom
8. laundry
9. artist studio

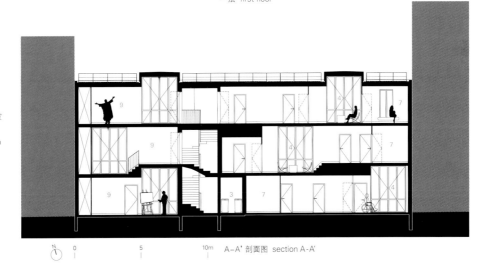

A-A' 剖面图　section A-A'

北立面_住宅B　north elevation_housing B

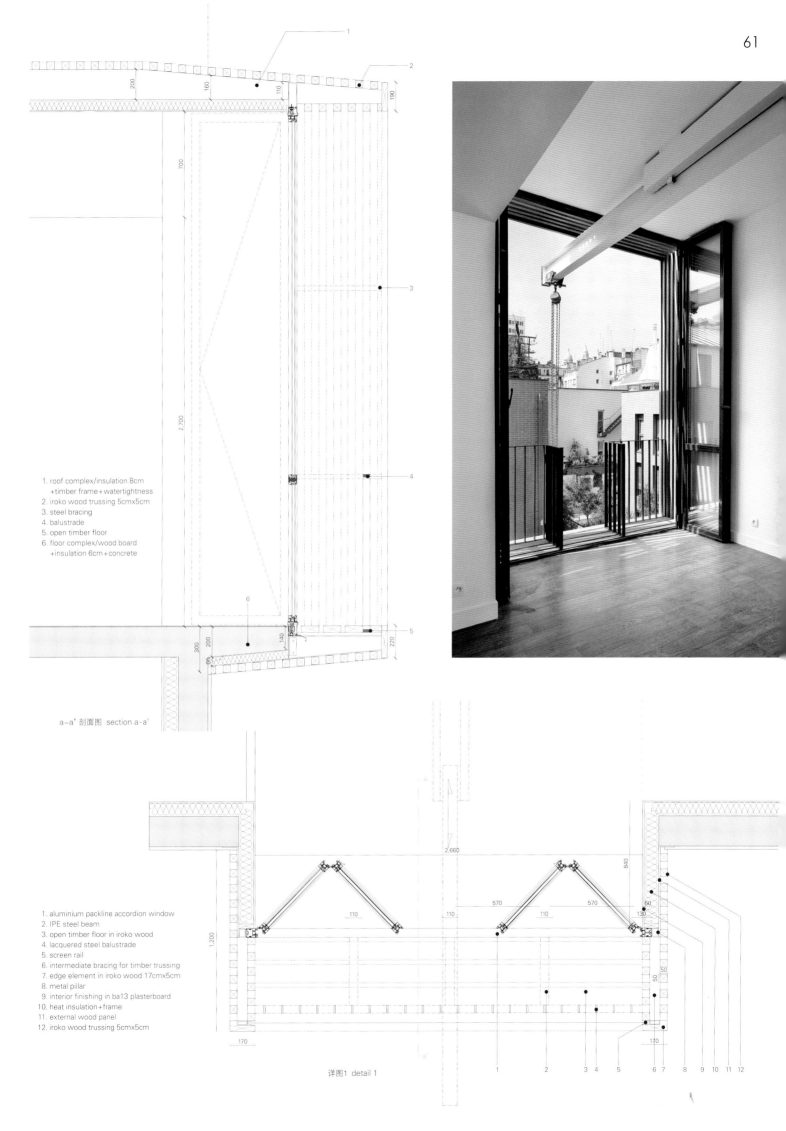

1. roof complex/insulation 8cm
   +timber frame+watertightness
2. iroko wood trussing 5cmx5cm
3. steel bracing
4. balustrade
5. open timber floor
6. floor complex/wood board
   +insulation 6cm+concrete

a-a' 剖面图 section a-a'

1. aluminium packline accordion window
2. IPE steel beam
3. open timber floor in iroko wood
4. lacquered steel balustrade
5. screen rail
6. intermediate bracing for timber trussing
7. edge element in iroko wood 17cmx5cm
8. metal pillar
9. interior finishing in ba13 plasterboard
10. heat insulation+frame
11. external wood panel
12. iroko wood trussing 5cmx5cm

详图1 detail 1

housing B, Rue du Nord

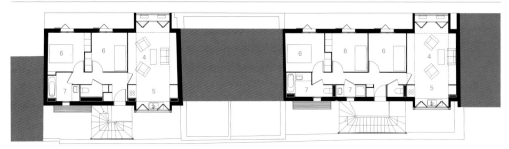

三层 third floor

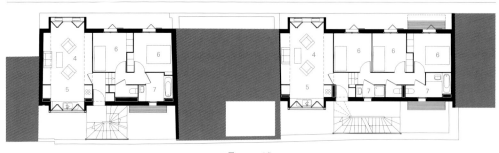

二层 second floor

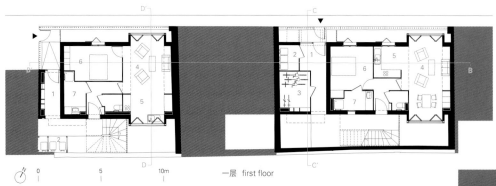

一层 first floor

1 大厅
2 垃圾室
3 自行车棚
4 起居室
5 厨房
6 卧室
7 浴室
8 洗衣房
9 艺术家工作室

1. lobby
2. garbage room
3. bike shed
4. living room
5. kitchen
6. bedroom
7. bathroom
8. laundry
9. artist studio

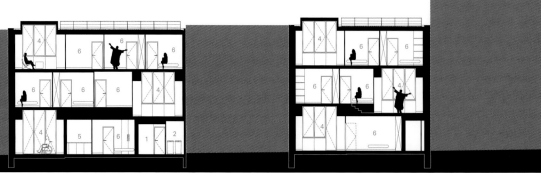

B-B' 剖面图 section B-B'

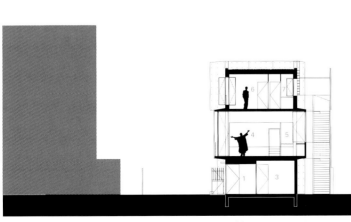

C-C' 剖面图 section C-C'

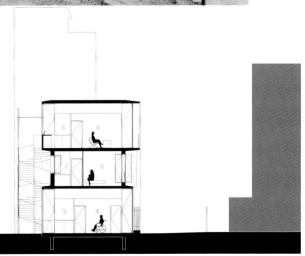

D-D' 剖面图 section D-D'

巴黎的30座社会住宅

KOZ Architectes

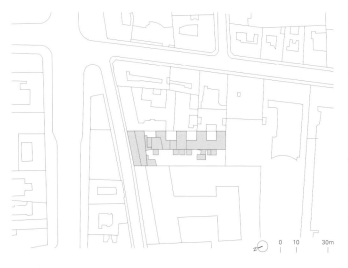

该项目坐落在巴黎北部一处狭窄的深处场地之内,这里是一处古老的工人阶层生活区。

现有的结构非常破旧。然后项目书要求对街道上的建筑加以保护,以保留其独特的精髓,同时,这里还存在着一种意识,即这里曾经发生了鼓舞人们的、富有的、密集的社交生活。

建筑师的首要目标是建造一座大型花园,向南方开放。它位于居民区的天然小径上,成为街道与个人住宅之间的中间空间。所有公寓的起居室都面向它开放,将其本身与这座城市中的小自然连接起来,同时也促进了邻里关系。

位于街道一侧的现有建筑全部进行翻新,以设置完全回应现今居住标准的公寓。该项目还建造了一个双高的门廊,允许人们从街道望向花园,或者对建筑再加以利用,在新建筑的建造期间沿着花园为人们提供入口。

新建部分的主要特点是100%为木结构,在其负责的部分,建筑师深信将材料与无法比拟的生态和美学效益相结合。如果一个国家所有的规章适度和建造商的习惯都能够围绕着混凝土施工来转的话,那便是一个了不起的成就。从地面到屋顶,建筑师必须想出新的解决方案,来解决结构、隔音以及消防问题。对技术级别较低的材料进行高科技的使用,将使整体的环境远远超过现时的标准。

尽管木建筑需要高级别的规章,但带有严格的设置节奏的插入式木箱构成的突出立面,扰乱了合理的建造次序,且赋予了场地一种自然的特点。这些木箱的随意位置使每个公寓的布局都是独一无二的,且可用作不同的用途,从官方的卧室,到SOHO空间,再到健身房。建筑后身的小庭院为一层提供了一个私人花园,且为所有的浴室引进自然光线,来增添舒适性,提高家居式生活质量。

木质覆层的绘图式布局使建成体量给人一种零碎的感觉,产生了一种安静的氛围,并且展现了粗糙的天然木的感官价值。它有助于将建筑与花园及其周围起伏的木质小径和平台相结合。这些小径和平台"漂浮"在小树以及一片野花之上。这使花园成为一间开放的房间,欢迎小团体住户的到来。它完全地展现了社会住宅能够成为这样的一个场地,即促进小规模的乌托邦区域的产生,人们生活幸福,且引以为豪。

30 Social Housing in Paris

The project is located in an old working class borough in the north of Paris on a deep and narrow plot.

Existing constructions were in a very poor condition. Yet the brief asked for the preservation of the building on the street to retain its picturesque spirit, and there was a feeling that there had been a rich and dense social life here that inspired people.

The architects' first objective was to create a generous garden, opened to the south. It is on the natural path of the inhabitants and acts like an intermediate space between the street and one's home. All the apartments' living rooms are opened on it, to connect oneself to this small piece of urban nature and to connect neighbors together.

The existing building on the street side is totally revamped to accommodate apartments answering today's living standards. It also wins a double height porch to allow views to the garden from the street, and to provide access during the construction of the new building along the garden.

The main feature of this new part is to be 100% in wooden construction. A strong conviction on the architects' part is that this material combines incomparable ecological and aesthetic benefits. But quite a feat in a country was all regulations and builder habits revolve around concrete construction! From ground to roof they had to invent new solutions to solve structural, acoustic or

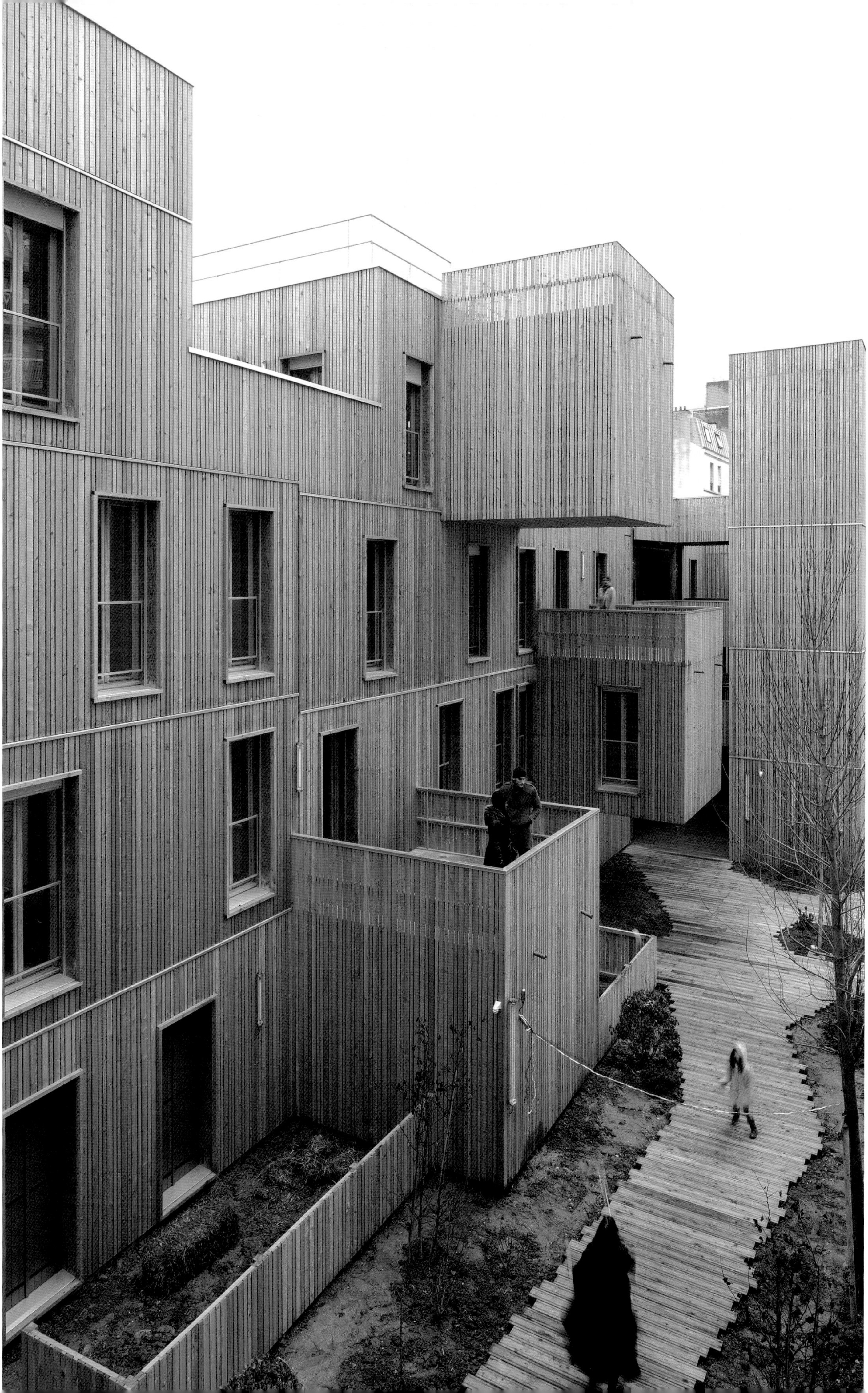

fire issues. A sort of high-tech use of a low tech material, brings the overall environmental performance to exceed by far current standards.

But even if wood building requires a high degree of discipline and stringency of the plug-in wood boxes with playful rhythm that punctuates the facade disrupt the rational constructive order, and give the place a spontaneous character. The boxes' aleatory positions make the layout of each apartment unique and allow for different uses, from their official bedroom function to SOHO spaces or gym room. Small courtyards at the rear of the building provide private gardens at the ground floor and help to bring natural light in all the bathrooms to add extra comfort and a house-like quality of life.

The very graphic laying of the wooden cladding further fragments the perception of the built volume, gives a very quiet tone and values the sensuous presence of the rough natural wood. It helps to blend architecture with the garden and it's undulating wooden path and terrace, floating around delicate trees and a prairie of wild flowers, thus making the garden a sort of open air room, welcoming to the small community of inhabitants. It ultimately states that social housing can still be the place to promote small scale sensitive utopias of well being and pride. KOZ Architectes

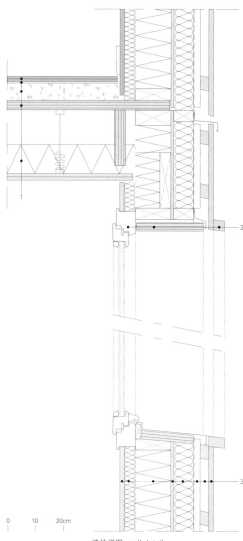

墙体详图 wall detail

1. wood flooring:
   fermacell panels 2x10mm, 10mm insulation domisol, granules fermacell 60mm, 37mm kerto, chord kerto S 45mm mineral wool 10cm ceiling on anti vibration suspension-2 BA13
2. frames: wood windows, frame ext. Q 33mm brown T.A.C, v. annodise, coaching larch Cl.3
3. wood frame wall:
   BA18, 4cm mineral wool, wool 13cm, OSB 12mm, 80mm rock wool, rain visor, cleats v. 45mm, taxxeaux h. 60mm, 20mm larch siding poses a clerestory.

1. smooth solid wood chaining
2. moise kerto Q 39mm
3. complementary rib kerto Q 39mm
4. reinforcement kerto Q 45mm
5. joint cover kerto Q 27mm bracing
6. lintel strip glue
7. post support pin header
8. complementary rib kerto Q 39mm
9. integrated fitting in frame transfer vertical forces

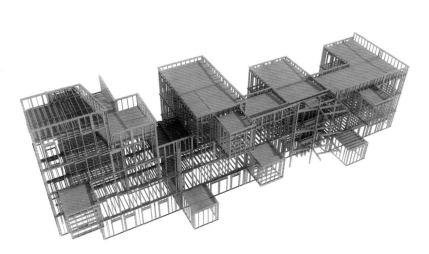

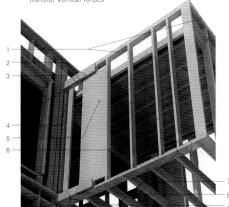

木框架详图 wood frame detail

A-A' 剖面图 section A-A'

项目名称：30 Social Housing in Paris
地点：94 Rue Philippe de Girard, Paris, France
建筑师：KOZ Architectes
项目团队：Collectif PAJE, EVP Ingnieurs Structure, DELTA Fluides, RPO Economie, PEUTZ Acoustique
甲方：SIEMP
规划：housing, commercial, parking
用地面积：1,261m²　总建筑面积：740m²
有效楼层面积：2,600m²
设计时间：2007　施工时间：2010—2012　竣工时间：2013
摄影师：©Cécile Septet (courtesy of the architect)

B-B' 剖面图 section B-B'

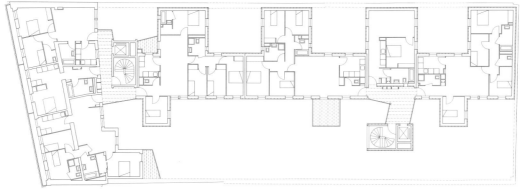

四层 fourth floor

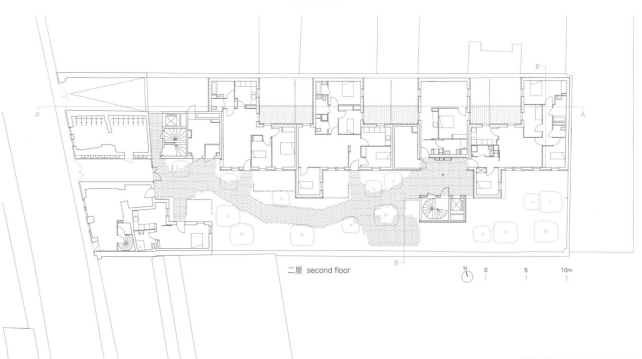

二层 second floor

圣玛丽亚住宅

Hierve

圣玛丽亚住宅位于Valle de Bravo中心的一处受保护的历史场地内，Valle de Bravo在16世纪30年代是一座小型殖民城市，离墨西哥城仅有两小时的车程距离。

这座历史城镇的地理环境非常优越，处于人工湖的外围区。项目场地距离圣玛利亚教堂100码，该教堂是一个殖民教堂，其历史可以追溯到16世纪。

该项目被构想成为墨西哥城繁忙的生活之后的周末休闲去处。它包括9栋独栋住宅和一些设施，提供了酒店般的体验。

面对着这样优美且幽静的环境，建筑师对此只带有尊重的情感，他们细心谨慎，尽力通过对地理方面的处理来产生最好的环境。地方当局在大多方面都给予了建筑师一些规定。

在设计的初期阶段，建筑师便试图与地形、现有植被、湖景、邻近建筑以及街道宽度等相呼应，使之与周围环境强烈地连为一体。

该项目的主要空间布局包括将体量沿着两条街道设置，街道对场地边界进行限制（这属于当地的规章制度），以及将住宅布局设置成L形。通过这种方式，建筑师能够充分利用朝向和视野的优势，同时能够产生一处建造水池的区域（就露台的意义而言），如同Valle de Bravo市中心的传统住宅一般。

该项目的大致布局包括两个楼层：

低层包括入口、停车场、保安室、垃圾房、小型自助餐厅、礼宾服务室、会计室、36个小酒窖、洗衣房、浴室以及维修间。

上层则包含一个沿着住宅延伸的独立走廊（L形）。这条走廊的建造目的是使其成为9个住宅和水池区域的缓冲区域。水池区域是一处开放且独立的空间，使房屋具有隐私感，但是又面向周围的景观开放。这个区域包括一个木质平台、一个按摩浴缸、两个水池、一处生火区以及一座小花园。每个独立走廊的尽头都有一个小天井，为瑜伽室和按摩室带来阳光和宁静。

每栋住宅（除了5号住宅之外）的低层都包括一个小型大厅、3间卧室和两间浴室。每栋住宅的上层则利用高度可以调节的优势，带来开阔的空间，为人们尽可能地提供湖泊的美景和周围的环境视野。这一层设有一个开放式厨房、起居室、一个私人阳台、一间小浴室以及一个小型洗衣间。

选择材料的出发点来自于一个相关的规定，该规定认为所有的外墙都应该使用白色抹灰饰面。在这样的一个中性色调且优美的背景下，建筑师决定将墙体与两种材料结合：天然石材和实木。

Santa Maria Housing

Santa Maria is a housing development located in a historic protected site in the heart of Valle de Bravo, a small colonial city dating from 1530 which is 2 hours away from Mexico City.

This historic town has a strong physical context and is found in the outskirts of a man-made lake. The site is located in a hundred yards from the church of Santa Maria Ahuacatlan, a colonial church that dates back to the XVI century.

The project is conceived as a weekend retreat from Mexico City's

项目名称: Santa Maria 地点: Valle de Bravo, Mexico
建筑师: Hierve
合伙人: Alejandro Villarreal 项目建筑师: Andrés Casares
合作商: Sugey Ramirez, Gabriela Rosas, Jesús Ramirez,
Denisse Novelo, Arturo García Crespo
结构工程师: Moncad 机械工程师: M3 Ingeniería Integral
灯光工程师: LLC Iluminación 景观工程师: Ambiente Arquitectos
内部设计: Isabel Maldonado 木工: Maderaje Arquitectónico
承包商: Zimbra 甲方: Inmobiliaria Sanmo SA de CV
用地面积: 2,509m² 总建筑面积: 2,269m²
造价: USD 2,700,000 施工时间: 2008—2010
摄影师: Courtesy of the architect-p.82, p.83
©Fernando Cordero (courtesy of the architect)-p.72~73, p.75, p.76, p.77, p.78

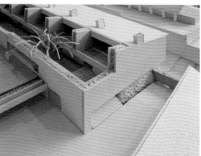
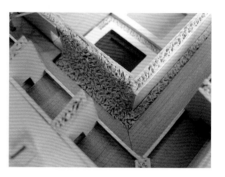
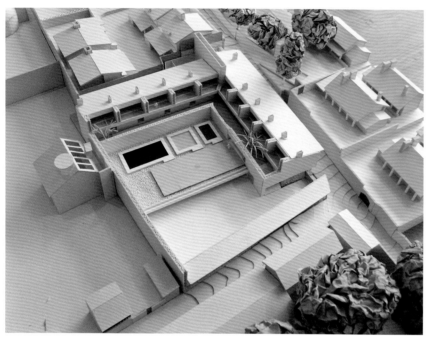
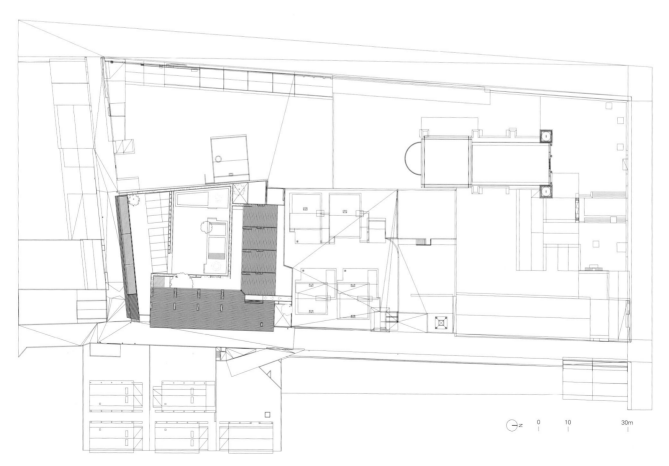

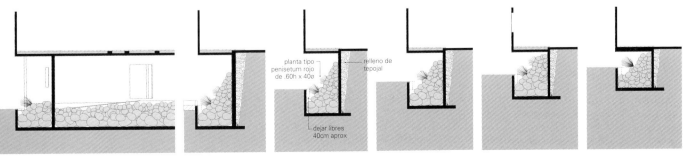

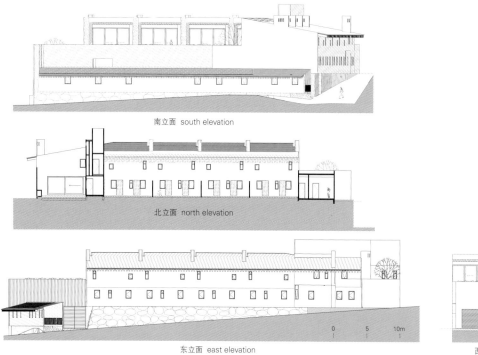

南立面 south elevation

北立面 north elevation

东立面 east elevation

西立面 west elevation

A-A' 剖面图 section A-A'

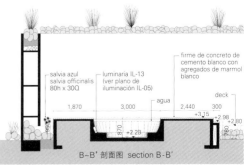

B-B' 剖面图 section B-B'

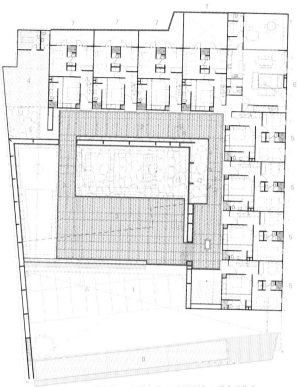

1 停车场 2 主要走廊 3 水池区域 4 瑜伽工作室
5 住宅类型-1 6 住宅类型-2 7 住宅类型-3 8 太阳能嵌板
1. parking 2. main corridor 3. pool area 4. yoga studio
5. house type-1 6. house type-2 7. house type-3 8. solor panels

一层 first floor

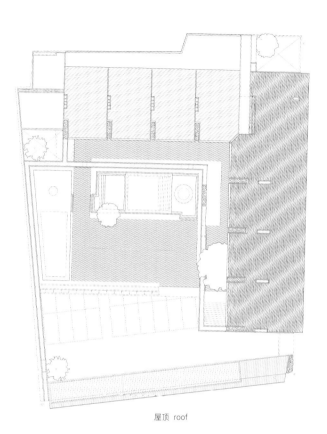

屋顶 roof

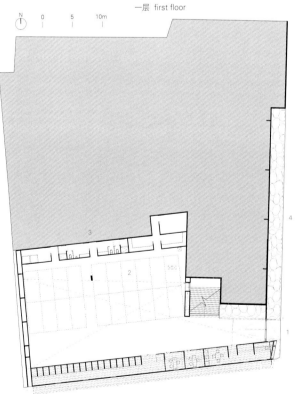

1 入口 2 停车场 3 服务区 4 沟渠
1. access 2. parking 3. services 4. ditch

地下一层 first floor below ground

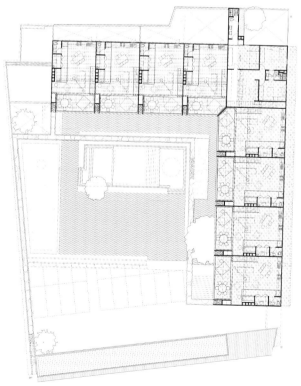

二层 second floor

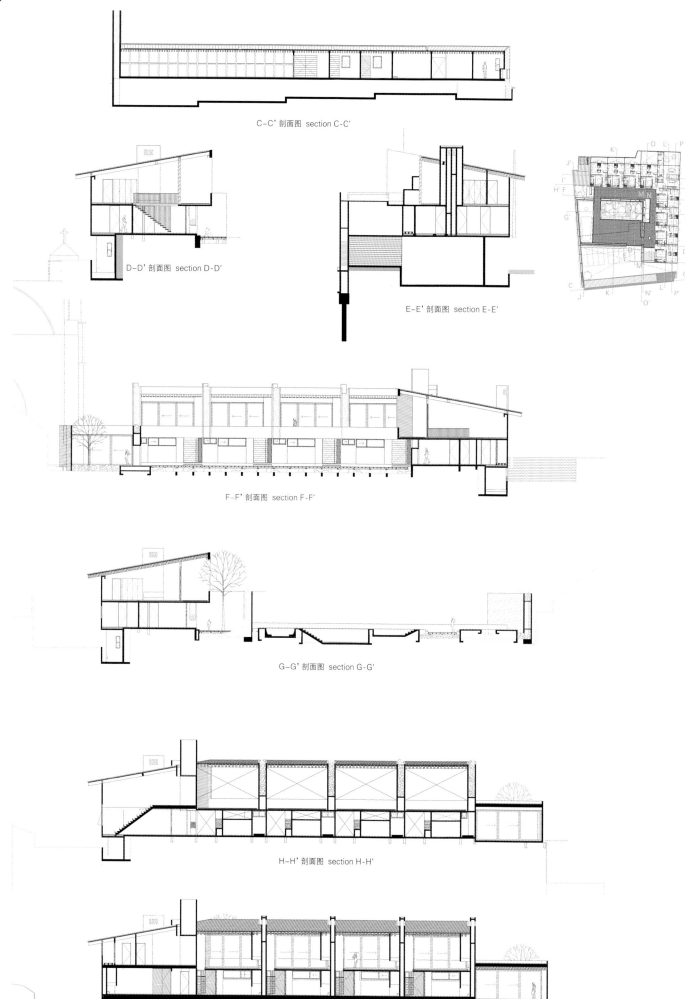

C-C'剖面图 section C-C'

D-D'剖面图 section D-D'

E-E'剖面图 section E-E'

F-F'剖面图 section F-F'

G-G'剖面图 section G-G'

H-H'剖面图 section H-H'

I-I'剖面图 section I-I'

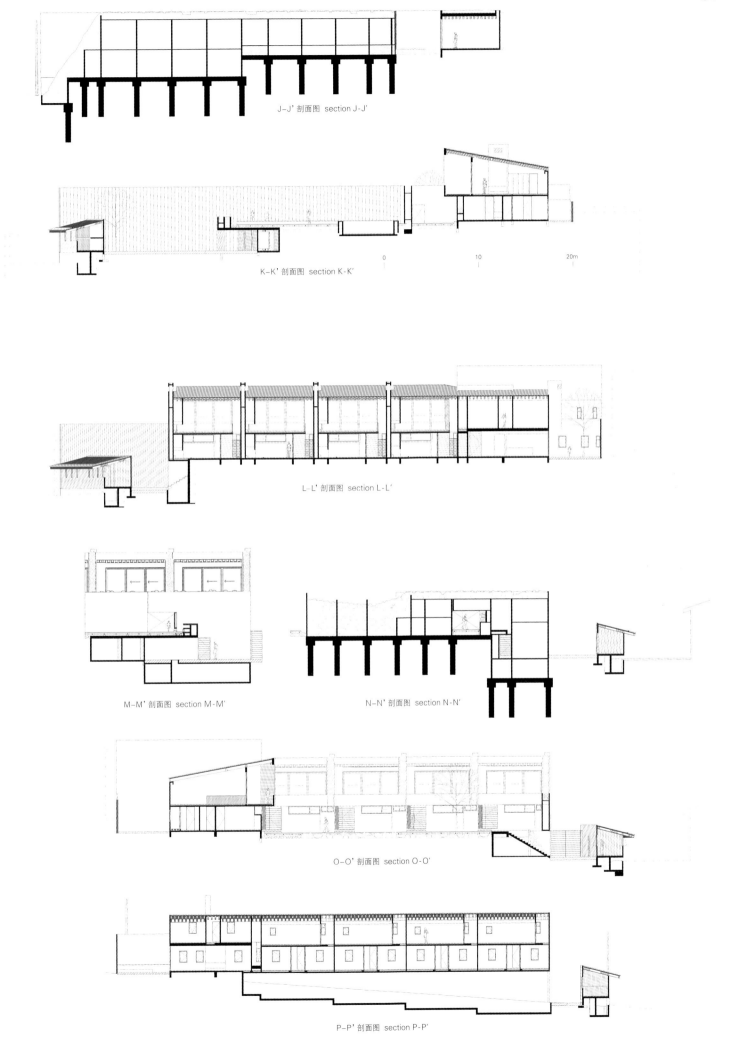

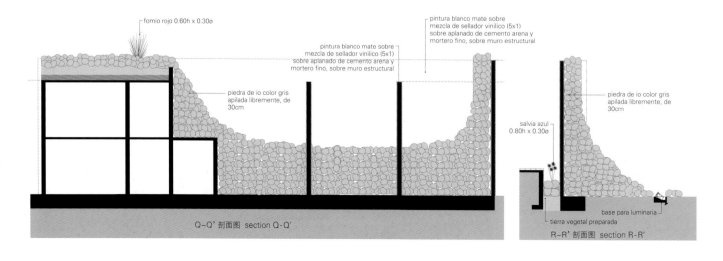

Q-Q' 剖面图 section Q-Q'   R-R' 剖面图 section R-R'

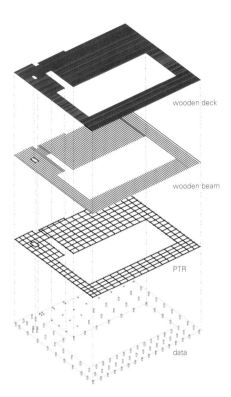

busy lifestyle and it includes 9 town houses and some amenities that provide an almost hotel-like experience.

When confronted with such a beautiful and restricted context, the architects could do nothing but respect it, trying to bring the most out of it by handling the physical aspects of the project with care. Most of this aspects were given to them as regulations by the local authority.

From the very early stages of design, the architects tried to make a strong connection with the site by responding to its topography, to the existing vegetation, to the views of the lake, to the neighboring buildings, to the width of the streets, etc..

The main spatial configuration of the project consists on placing volumes along the two streets that confine the site (which is a local regulation), and placing the houses in an L-shape. By this, the architects could take advantage of the orientation and the views, and at the same time they were able to generate a pool area (in the sense of a main patio), just like in the traditional houses found in downtown Valle de Bravo.

The general layout of the project includes two main levels:

The lower one includes the entrance, parking space, security guard room, garbage room, a small service cafeteria, concierge room, accounting room, 36 small cellars, laundry, bathrooms and maintenance rooms.

The upper level includes a self contained corridor (in L-shape) that runs all along the houses. The purpose of this corridor is to become a buffer between the 9 houses and the pool area. The pool area works both as an open and a contained space, having a sense of privacy towards the houses, but opening up to the surrounding landscape. This area includes a wooden deck, a jacuzzi, two pools, a space for making fires and a small garden. At the very end of the self-contained corridor, there is a small patio that brings light and calm to a yoga studio and a massage room.

The lower floor of each house (except house number 5), includes a small lobby, 3 bedrooms and two bathrooms. The upper floor of each house takes advantage of the height regulation, bringing up an expansive space that offers great views towards the lake and the surroundings. On this floor there is an open kitchen, a living room, a private terrace, a small bathroom and a laundry closet.

The starting point for choosing the materials came from a regulation that considers that all the exterior walls should have a white plaster finish. Having such a neutral and beautiful background, the architects decided to combine it with two more materials: natural stone and solid wood. Hierve

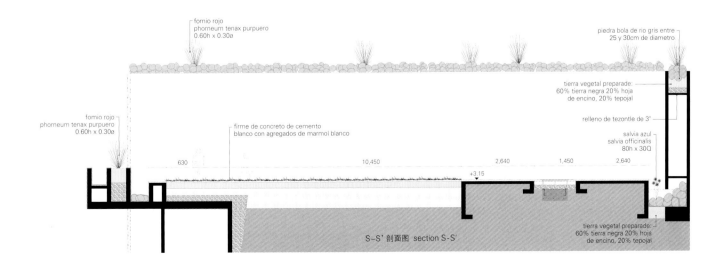

Plantoun的小村庄

Agence Bernard Bühler

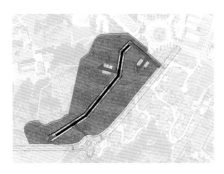

一个建筑项目旨在回答如下问题：如何在此地生活，要以怎样的形式存在于此地。由于场地较为特殊且罕有，所以这个项目将会促进这些问题的答案的产生。该项目是一处共享的公共空间，充满了亲密的氛围。环境产生了许多双重性，即致密与保护，私有化与渗透化。建筑师要面临的重要选择包括空间利用、项目进程、可视性以及亲密性的优化程度。建筑目标即存在于建筑背景中。

这个拥有39个单元的"Hameau de Plantoun"住宅距离著名的由Marcel Breuer于1958年至1961年建成的"Hauts de Sainte-Croix"优先城市化区（ZUP）不是很远，建在一条蜿蜒道路的两侧，与小径，即宽敞的斜坡式木质场地连接起来。人们能够从Marcel Breuer大道进入到分割的建筑中，这些被分割的建筑包括12座T3、17座T4以及10座T5，类型多样。建筑师所采用的结构规划——将木材作为主材料——与由住宅选择所引起的日常管理费用减少的需求（由于预制过程和标准化的实施）以及环境问题息息相关。轻质的木质材料突出了微型桩系统，使位于先天不利的地面上的住宅结构更加便于人们使用，且促进了其与木质场地的融合。

所有南侧的住宅都有一个围栏围起来的种植花园，面向森林开放。位于桩柱上的住宅均带有一个面向木质场地的露台。所有的住宅在四个方向都能受益。住宅被设计成拥有自身体量规格的简单标准模块，它们通过自身与景观和类型变化的关系来寻求独特性。这些位于传统区域的现代住宅确保买家负担得起。基本的模块式住宅（每座都是相同的）的面积为6.45m×6.45m，在布局方面具有灵活性，是真正具有可伸缩性的住宅。

由木质住宅构成的项目为环境建造了一座创新的建筑，满足了降低结构成本的需求。

因此，个人住宅选择所产生的额外成本由减少的预制和标准化成本来抵销，同时形成大面积的住宅群。此外，木材的轻质性与微型桩的使用相结合，使不利地形上的施工成为可能，同时使成本最小化，且与木质自然场地相融合。另外，通过使用木材，建筑的保温和隔音舒适性也能够达成，并且对较远环境所产生的影响也进行控制。

最后，具有独特的特点、现代化且创新的住宅远离周围经典的住宅环境，形成了一幅全新的社会住宅形象。

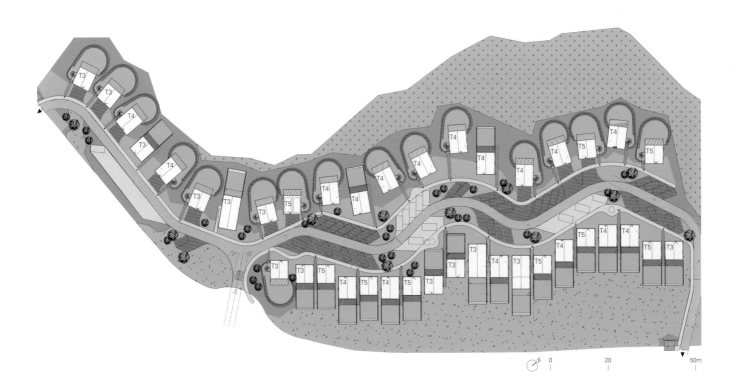

Small Village, Plantoun

An architectural project aims to answer the questions: how to live here, and how to be here. This program motivates these questions especially since the site is special and rare. It is a shared public space, which will accommodate intimacies. The environment generates many dualities like densify and preserve, privatize and permeabilize. Important choices include the use of space, the course, the visibility, prioritization degrees of intimacy. The architectural object is in the background.

Designed not far from the famous ZUP(Priority Urban Area) the "Hauts de Sainte-Croix", built by Marcel Breuer in 1958~1961, the 39 units of "Hameau de Plantoun" are distributed on either side of a serpentine road, associated with paths, a steep wooded site and generously. The subdivision, accessible from the Avenue Marcel Breuer, has 12 T3, 17 T4 and 10 T5, responding to a variety of types. The structural scheme adopted by the architects – the use of wood as the main material – is bound by both the need to reduce the overhead caused by the choice of houses, through the imple-

mentation process of prefabrication and standardization, but also environmental concerns. The lightweight wood, which reinforces a system of micro-piles, has also facilitated the construction of houses on this land which was priori unfavorable, and promoted their integration into the natural wooded site.

All houses in the south have a fenced and planted garden that opens to the forest. The homes on stilts have a terrace to the wooded area. All the houses benefit from four directions. The houses are designed from simple standard modules in their volume, and they find their individuality through their relationship with the landscape and changes in typology. These are contemporary houses in "traditional" areas to ensure affordable to buyers. The basic module house(same for each) has dimension of 6.45m× 6.45m, it allows flexibility in the distribution and is a real scalability housing.

The project construction of wooden houses, meets the need to reduce construction costs by offering an innovative architecture to the environmental dimension.

Thus, the additional cost generated by the choice of individual housing was offset by prefabrication and standardization while offering a wide range cluster. In addition, the lightness of the wood, combined with the use of micro-piles, makes possible the construction of a priori unfavorable terrain, minimizing costs and enabling integration into the natural wooded site. In addition, a thermal and acoustic comfort is made possible by the constructive solution of wood, limiting further environmental impacts.

Finally, the singular, contemporary and innovative housing is detached from the classic habitat surrounding, offering a new image of social housing.

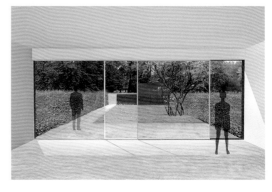

项目名称：Hameau de Plantoun
地点：Avenue Marcel Breuer, Bayonne, France
建筑师：Agence Bernard Bühler
项目团队：Agence Bernard Bühler(mandataire), N°5 Marie Bühler Architecte(association)
工程顾问：EGIS Sud-Ouest
甲方：Ophlm De Bayonne
用地面积：3,417m²
总建筑面积：17,751m²
有效楼层面积：7,434m²
设计时间：2006.9—2007.9
竣工时间：2009.5
摄影师：©Vincent Monthiers (courtesy of the architect)

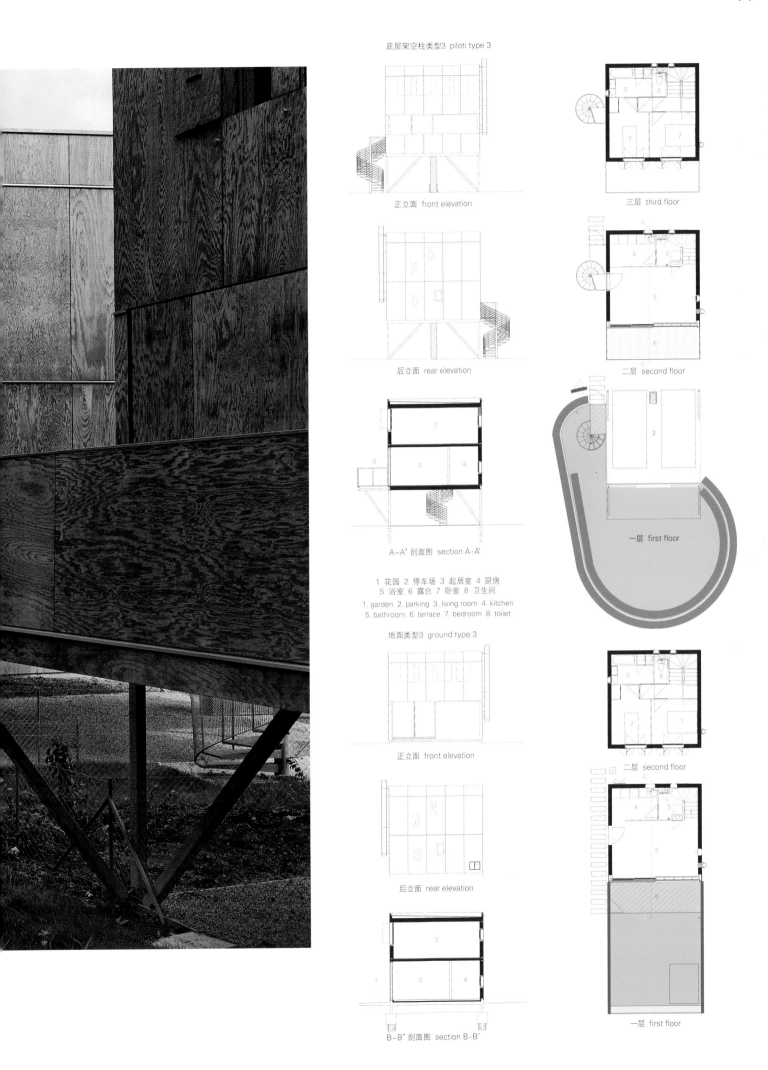

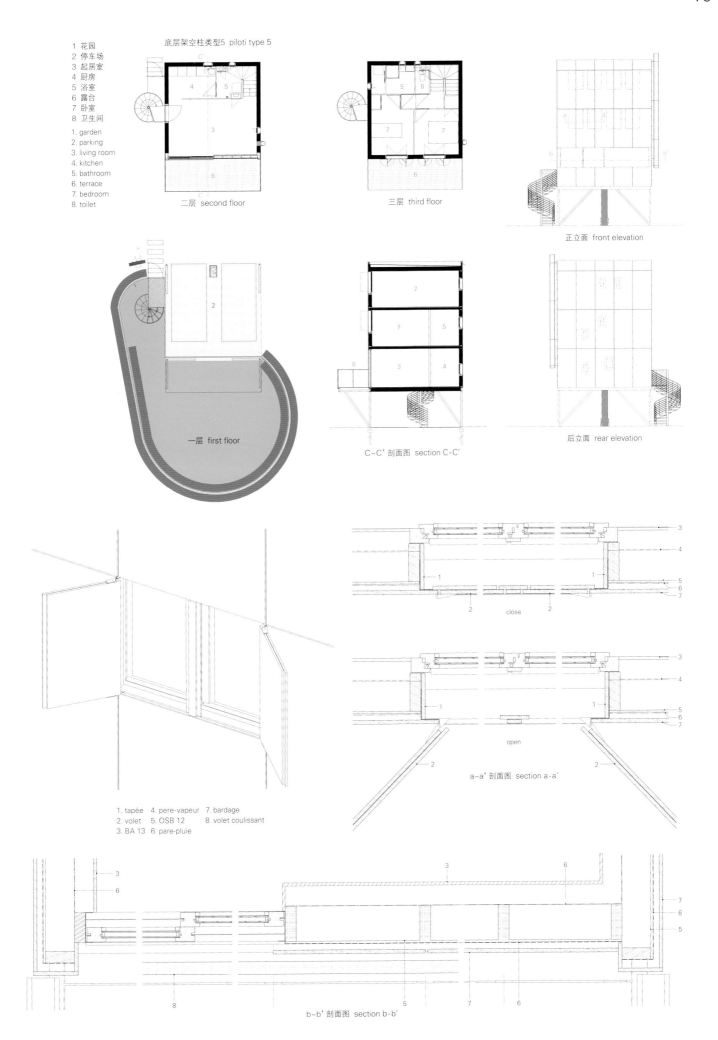

44个社会住宅单元

LEM+ Architectes

项目名称：44 HQE social housing units
地点：Marnes-la-Coquette, France
建筑师：LEM+ Architectes
工程师：Scoping
承包商：TMCR
甲方：Immobilière 3F
表面面积：3,338m² 总建筑面积：2,027m²
造价：EUR 6,209,673
设计时间：2007 竣工时间：2013
摄影师：©Julien Lanoo (courtesy of the architect)

该项目地处圣克卢国家公园的边缘处，由建筑师Pierre Lépinay和Bertrand Meurice设计，这个项目是为I3F的社会住宅发展项目组而设计的，位于Garches-Marnes火车站和A13高速公路围合的绿化带区域内，十分独特。

该项目（3338m²的平面区域）包含铁路一侧的带有私人花园的独栋住宅，且通过一组带有阳台的公寓（从工作室公寓到带有四至五个房间的复式单元），来呈现一个延长的U形布局，以保护中心的建筑免受噪音的骚扰。这些噪音都来自于南部的高速公路。这些住宅单元设有大量的窗户，因此产生了双倍的效果，房间特别明亮。白天，通过特殊设计的屋顶轮廓来对阴影面积进行限制，使私人花园沐浴在阳光下。

建筑师提供了一系列的创新方案，来克服噪音的限制问题。除了具有保护性的U形布局之外，他们还沿着高速公路一侧设置了木质围栏。建成的二层建筑带有加强绝缘系统（设置在阁楼空间），安装了带有连续的9m高脊线的斜坡屋顶，并且在高速公路一侧的住宅单元中配备了隔音门廊。

作为住宅单元的一处延展的绿化区，这个隔音门廊使公寓面向南方开放，并且提供了大面积的朝向圣克卢国家公园的玻璃空间，同时保护居民免受噪音的打扰。种植区面向天空开放，可由玻璃屏风围合关闭，以形成一处特殊的、有效的隔音屏障，这一屏障在公寓的特殊布局的作用下，隔音效果得以加强。杂物房面向高速公路开放，同时受保护的一侧的主要房间面向阳台、门廊以及私人花园开放。噪音通过双流系统得以减少，这个双流系统可以避免声桥的产生，且能够提高空气质量，促进住宅单元内的供暖。

怀着使用节能方案的愿望，建筑师选择使用带有隔热系统（位于突缘饰板）的室内保温系统，在屋顶安装了太阳能嵌板，并且为每座住宅单元都配备的太阳能集水器水槽。建筑师还特别关注了景观工程。攀爬植物种植在场地周围，在铁路沿线的围栏以及高速公路一侧150m长的支撑墙上生长。私人花园位于建筑体块的中央，被U形布局大致划出轮廓，其规模尽可能地减小，以提供更大面积的公用空间。公用空间种有不同的植物物种，由业主细心打理。

这个住宅项目见证了LEM+建筑师事务所发明的一种交替性景观类型。这种类型致力于使住户幸福安康。经过近些日子的交房，一半以上的住宅单元都住进了用户。

44 Social Housing Units

Located on the edge of the Saint-Cloud national park, the project by architects Pierre Lépinay and Bertrand Meurice for the I3F social housing development group is set in a particularly green area enclosed by the Garches-Marnes railway station and the A13 motorway.

Comprising single-family homes with private gardens on the railway station side and by a group of apartments provided with

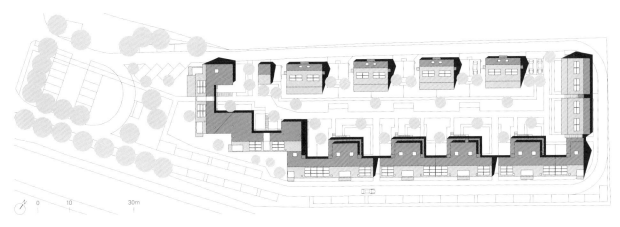

balconies ranging from studio flats to four and five room duplex units, the operation (3,338m² net plan area) takes the form of a drawn-out U-shape that protects the center of the block from the noise nuisances resulting from the presence of the motorway to the south. These housing units with their large number of windows and dual aspects are particularly bright. The private gardens are bathed in sunlight throughout most of the day thanks to roof profiles specifically designed to limit shadow impact.

The architects provided a number of innovative solutions to overcome the noise constraints. In addition to the protective U-shaped layout, they incorporated a wooden fence running alongside the motorway, built two-story buildings with reinforced insulation in the attic spaces, installed sloped roofs with a continuous nine meter high ridge line, and equipped the housing units on the motorway side with acoustic loggias.

Acting as a green extension to the housing units, the acoustic loggias open the apartments onto the south, providing large glazed spaces giving onto the Saint Cloud Park while also protecting residents from noise. This planted space, opening onto the sky and closed off to the sides by glass screens, forms a particularly efficient acoustic barrier that is further reinforced by the specific layout of the apartments. The utility rooms give onto the motorway side while the main rooms open onto balconies, loggias and private gardens on the protected side. The noise nuisances are also reduced by the use of a double flow system that avoids the creation of acoustic bridges, improves air quality and contributes to heating the housing units.

Wishing to adopt energy-saving solutions, the architects chose to use interior insulation with thermal breaks on slab nosings, install solar panels on the roof and equip each housing unit with solar collector tanks. They also paid particular attention to the landscaping project. Climbing plants have been placed around the site to colonise the fence on the railway line side and the 150 meter long support wall on the motorway side. Lying in the heart of the block outlined by the U-shaped layout, the private gardens are reduced in size as much as possible to offer larger spaces to the common areas that are planted with a variety of species and looked after by the landlord.

This housing project has seen that the LEM + architects invent an alternative type of landscape that contributes to the wellbeing of the residents. Recently handed over, over half of the housing units are already occupied.

1 露台
2 起居室
3 门廊
4 阁楼

1. terrace
2. living room
3. loggia
4. attic

A-A' 剖面图 section A-A'

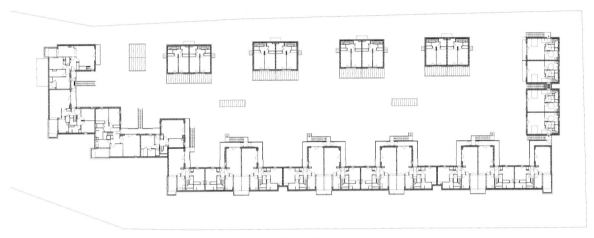

二层 second floor

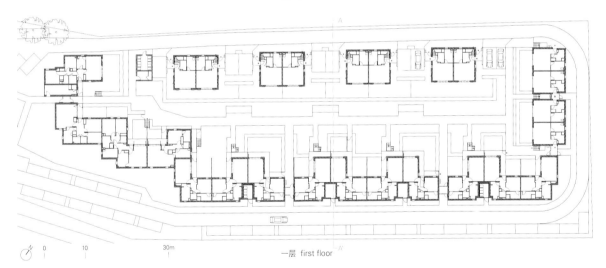

一层 first floor

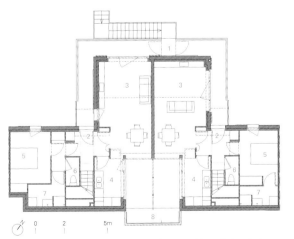

1 露台 2 入口 3 起居室 4 厨房 5 卧室 6 卫生间 7 浴室 8 门廊
1. terrace 2. entrance 3. living room 4. kitchen 5. bedroom 6. toilet 7. bathroom 8. loggia
复式单元_一层 duplex unit_first floor

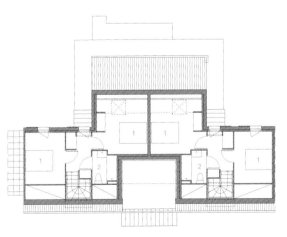

1 卧室 2 浴室
1. bedroom 2. bathroom
复式单元_二层 duplex unit_second floor

中心村落

5468796 Architecture + Cohlmeyer Architecture Limited

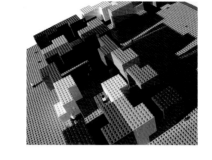

温尼伯的中心村落住宅以为贫困家庭服务为目的，利用设计来使被忽视的市中心贫民区恢复活力，并且为住户提供独特的环境，以激发自豪感，鼓励社区建筑的开发。

场地是一处废弃的L形区域，这是6个家庭独栋住宅的所在地。然而，该项目建立了一个微型村庄，在6座三层住宅建筑（很容易地建造和保留）内设置了25个寓所。建筑的布局划出了两处公共空间，并且使其富有生气。这两处公共空间分别为一条直通街道以及一个共享庭院，通过这个项目在城市中"迂回"，并且为住户和周围邻里提供公共设施。风景优美的庭院为儿童玩耍提供了一处安静且安全的区域，而新街道则成为一处非正式的会见场所。每个寓所都有自己的入口，或者在地面层，或者在室外台阶上，以减少室内交通流线的设置以及整体建筑的规模，并且促进住户之间的联系，鼓励他们的交往。

为了使空间最大化，材料与光线得以最大程度的利用，每个单元是由较为紧凑的8′×12′（2.44m×3.66m）的模块以及14′×12′（3.66m×4.27m）的悬挑模块构成，这些丰富且有趣的构成元素应用在大型起居区域，以提供更广阔的视野。这些模块堆叠在一起，相互联结，以创造不同的单元布局，包括34.8m²的单间卧室单元和81.3m²的四间卧室单元。每个单元的卧室都分布在不同的楼层上，即使是小型公寓也有足够的私隐空间，这对于大量的用户是多代家庭来说，是一个十分重要的因素。

入口和窗口在建筑的所有面上都有所设置，因此，整个项目并没有后面，提高了街道质量，使建筑在所有方向都具有安全感。此外，一座典型的住宅至少要在两侧设置八个甚至更多个窗户，来为视野、光线和对流通风提供足够且不同的入口。橙色的凹陷形整流罩设置在窗户的周围，用来调节隐私度和进入单元内的视线，赋予中心村落一个独特的城市身份。随着时间的流逝，这个项目正在证明富有创意的建筑是怎样影响城市主义，以及对社会关系产生积极影响的。

可持续性

中心村落的主要的可持续倡议是在既有的场地上，在三层联体别墅类型建筑中实现单元的密度。然而，如果该方案不适宜居住，或者让步于空间质量，那么，实现这种单元的密度便不算是创新的设计。原始的六栋独栋住宅进行合并，以建造容纳25个寓所的场地，住宅的数量是由甲方来决定的。四倍的密度使场地有可能被填满，没有共享的社区空间、住宅单元的呼吸空间，甚至没有了设置窗户的墙体区域，这将会使室内和室外空间都无法居住。

为了充分利用潜在的建筑体块，建筑师决定设计房间模块，包括规为8′×12′（2.44m×3.66m）的厨房、次卧、垂直流线区以及洗手间，而悬挑出来的模块的规格为14′×12′（2.44m×3.66m），用于大型起居空间和主卧。将这些模块进行仔细的结合，以产生一系列的单元，用于个人和家庭居住，同时证明场地内建有较少的体块，同样可以容纳所需的单元数量。

Center Village

Serving underprivileged families, Winnipeg's Center Village housing cooperative utilizes design to help revitalize a neglected inner-city neighborhood and to provide its residents with a unique setting that inspires pride and encourages community-building. The site was an abandoned L-shaped lot zoned for six single-family houses. Instead, the project established a micro village of 25-dwellings within six, three-story blocks that would be easy to build and maintain. The blocks' arrangement both defines and animates two public spaces – a through-street and a shared courtyard – that weave the city through the project and provide amenities for residents and the surrounding neighborhood. The landscaped courtyard offers a calm and protected place for children to play, and the new street is an informal meeting place. Each dwelling has its own entrance, either at grade or on an exterior staircase, thus reducing internal circulation and the size of the overall building, and also prompting residents to connect and get to know one another.

Designed to make the most of space, material and daylight, the units have rich and playful compositions made from compact 8'x12' modules and cantilevered 14'x12' modules for larger living areas that offer broader views. The modules are stacked and interlocked to create diverse unit configurations that vary from 375 square feet for a one-bedroom unit, to 875 square feet for four-bedroom units. Since the rooms of each unit are distributed over several floors, even small apartments have plenty of privacy, which was an important factor since a number of the tenants are multi-generational families.

With entries and windows positioned on all sides of the blocks, there is no rear side to this project, thus improving the street quality and safety and security in all directions. Further, a typical residence has eight or more windows on at least two sides of the building, providing ample and varied accesses to views, daylight and cross-ventilation. Deeply set, vibrant orange cowlings around the windows modulate privacy and views into the units, granting Center Village a distinct identity in the city. With time the project is demonstrating how inventive architecture can influence urbanism and positively impact social relationships.

Sustainability

The primary sustainable initiative of Center Village is the density of units achieved within a three story townhouse typology on the given site. Density on its own, however, is not innovative if the solution is not livable or compromises on the quality of space. What was originally six single family house lots were amalgamated to create the site for 25 homes, and the number of units was required by the client. Quadrupling the density had the potential to fill the entire lot, leaving no shared community space, breathing room for the units, or even enough wall area for windows which would have rendered both the interior and exterior spaces uninhabitable.

To combat the potential building mass the architects decided to design room modules consisting of 8'x12' for kitchens, secondary bedrooms, vertical circulation and washrooms, and cantilevered 14'x12' modules for larger living areas and master bedrooms. Careful mixing of these modules resulted in a range of unit sizes for individuals and families, and proved that less mass could be built on the site and still accommodate the required number of units.

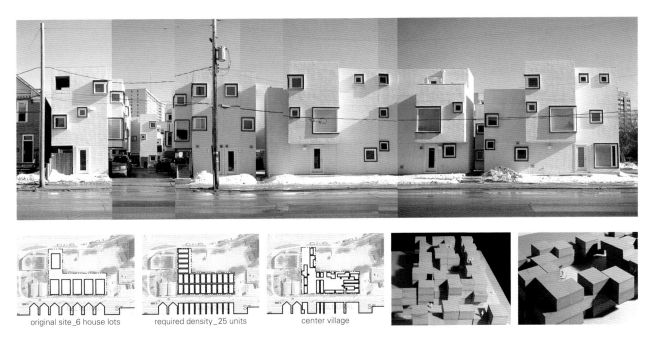

original site _ 6 house lots | required density _ 25 units | center village

项目名称：Center Village
地点：Winnipeg, Manitoba, Canada
建筑师：5468796 Architecture, Cohlmeyer Architecture Limited
项目经理：Hold Zone Inc.
结构工程师：Lavergne Draward & Associates
土木工程师：MEC Consulting
施工单位：Capstone Construction
景观建筑师：Cynthia Cohlmeyer Landscape Architect
甲方：Centre Venture Development Corporation
用地面积：1,362m²
总建筑面积：1,674m²
造价：CAD 2.5M
竣工时间：2010
摄影师：©James Brittain

建筑A_西立面 building A_west elevation

建筑A_北立面
building A_north elevation

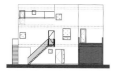
建筑A的内侧_东立面
building A inside_east elevation

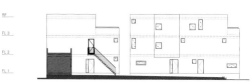
建筑A的内侧_西立面
building A inside_west elevation

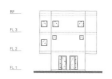
建筑B_西立面
building B_west elevation

建筑B_A-A'剖面图
building B_section A-A'

建筑B_B-B'剖面图
building B_section B-B'

建筑C_西立面
building C_west elevation

建筑C_C-C'剖面图
building C_section C-C'

建筑C_D-D'剖面图
building C_section D-D'

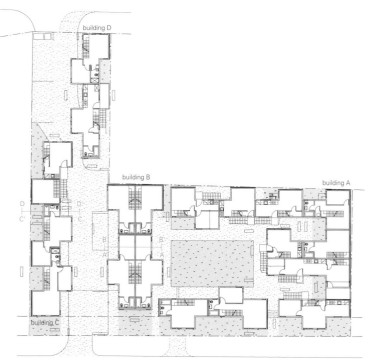
二层 second floor

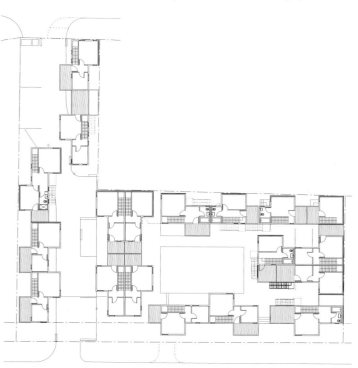
三层 third floor

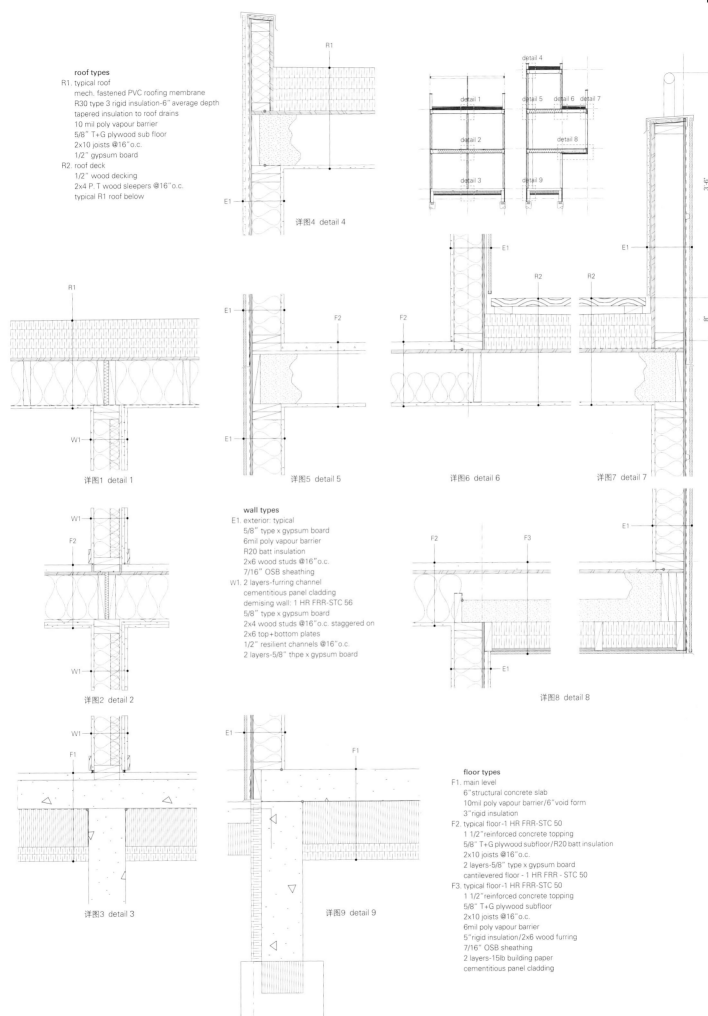

居住的流变 | Dwelling Shift

红苹果住宅公寓
Aedes Studio

红苹果住宅公寓周围的建筑大都可以追溯到20世纪70年代。这些建筑的规模较大，内部带有足够的空间和大量的绿地。由于整个场地区域是在短期内建成的，且建成不久，所以整体来看缺乏城市中心具有的典型的历史层次。在这里，与自然的连接是较为直接的，且这种连接并不少见。通向城市的入口十分便利，简洁易达，使这处区域十分适合居住。尽管缺乏历史和氛围，但是建筑师接下这一艰巨的任务，来设计一座全新的"旧建筑"，即一座带有历史的现代建筑，也是一座包含对立面的建筑。建筑师对这座建筑使用的手法仿佛他们要恢复将一座位于市内的古老废弃建筑的活力。他们将一座废弃的工厂建筑想象为起始点，经过翻修之后会成为一座豪华的、人们渴望居住的场地。然而，既然现实中并没有一座工厂来作为起始点，那么建筑师就必须去创造一座。这就是"充满活力的工厂"理念通过许多对立的理念（与"旧"和"新"相关），从而得以产生且被赋予特色的过程。这一理念的支撑点是砌砖——将新旧材料融为一体的材料。

结构（旧到新/从和谐秩序到动态混乱）

该建筑代码包含一个预制的砖薄壳，其轮廓特点与场地的不规则性相匹配，以形成能够突出透视场景并且成为重要特色的锐角。洞口几乎一致，排列顺序较为严格，这种顺序起源于砌砖的网格。在较为随意的地方或者不设置洞口，或者通过大型凸起（两层窗户）来重复设置。在平滑的墙体改变其结构的地方，每隔一块砖的一半也突出墙体表面。通过这种方式，悬挑出来的网格在第三维度中不断地变化。

树木出现在了室外墙体的大型凿洞中。它们位于一层的地面上，种在建筑的周围，并且持续地向上生长，延伸至屋顶。采用这种方式，自然所带来的影响力突出了过去的时代感，形成了浪漫的历史。

和普通的工厂一样，这座创造出来的工厂的屋顶也有很多烟囱。它们在建筑师的解释下成为采光井和盆景。在一层，这些元素也有所出现，打破了室内外之间的界限。另一方面，这些烟囱非常类似于立面上突出的砌砖，只不过是规格大一些而已。同样的规格差异也出现在两层的起居空间与常规的洞口中。

楼梯是整座建筑中最为夸张的空间，它成为一个大型采光井的代表，住宅的平台围绕楼梯来设置，它同立面一样，采用了同样的薄砖壳。

公寓里包含一个与立面距离较远、岛式的体量（房间）。因此，住户从室内就有可能注意到立面洞口的韵律。双高的起居室设计参考了纽约的阁楼设计。建筑师特意将其规模设置得较大，以突出从工业功能转为住宅功能的感觉。

这座建筑非常重要的特点是位于二层的运动场。建筑材料采用了与金属悬臂阳台相同的处理手法，但是又一次地被高度夸大。

Red Apple Apartments

The surrounding neighborhood consists mostly of apartment blocks that date back from the 1970's. The buildings are large with enough space in-between and plenty of greenery. Because the whole area is built in relatively short period of time and not very long ago, it lacks the typical historic layers of the city center. Here the connection to nature is direct enough, the access to all city – conveniences – is fast enough and easy, which makes the area nice to dwell. In spite of that it still lacks history, the past and atmosphere.

The architects set the uneasy task to design a new "old building"; a contemporary building with past – a building that contains opposites. They approached this project as if they had to revitalize an old, abandoned part of the city. They imagined as starting point an abandoned factory building, which after renovation becomes a luxury and desired place for habitation. But since they didn't have a factory to begin with, they had to create it. This is how

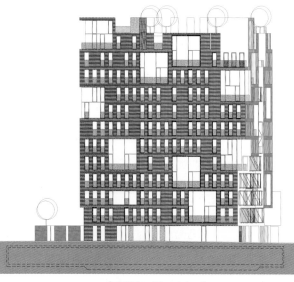

东北立面 north-east elevation

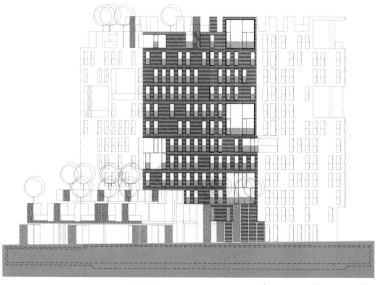

西南立面 south-west elevation

0　10　20m

the "living factory" arised, and is characterized by many opposing ideas related to the terms of "old" and "new". The backbone of the concept is the brick – the material that brings together the idea of old and new.

The structure (old – new/harmonious order – dynamic chaos)

The code of the building consists of a perforated brick shell. Its outline deliberately follows the irregularities of the site, creating acute angles which enhance the perspective and establish a significant character. The openings are completely similar, in strict order that originates from the brick's grid. In random places they are missing or are replaced by large break-through (two story windows). Where the smooth wall changes its structure, every second brick sticks halfway out of the wall surface. This way the cantilevers' grid evolves in the third dimension.

In large holes in the outer wall appear trees. They surround the building in the ground floor and keep growing onto it up to the roof. This way nature's influence enhances the feeling of the past time and the romantic "old".

As every ordinary "factory", this one as well has many chimneys on top. They again undergo the architects' interpretation and are used as light shafts or tree pots. In the ground floor this elements make an appearance, also breaking the line between inner and outer space. On the other hand they very much resemble the sticking-out bricks of the facade, but in bigger scale. Similar scale difference is found in the relation between the two-story living room-windows and the ordinary openings.

The staircase is the most highly hyperbolized space in the building. It represents a huge light shaft, around which the platforms to the dwellings are gathered. It consists of the same perforated brick shell as the facades.

The apartments contain island-like situated volumes (rooms) away from the facade. This makes it possible for the inhabitants to notice the rhythm of the facade-openings from the inside. The double height living rooms make a reference to the New York lofts. Their bigger scale is deliberate in order to enhance the feeling of the function which changes from industrial to residential.

Very important feature of the building is the sports ground on the second floor. Regarding the building materials – it is treated the same way as the metal cantilever balconies, but again highly hyperbolized. Aedes Studio

项目名称：Red Apple
地点：39 Petko Todorov Blvd, 1000, Sofia, Bulgaria
建筑师：Aedes Studio
甲方：sofbuild
功能：apartments, offices, stores
用地面积：2,640m² 总建筑面积：1,129m² 有效楼层面积：11,000m²
竣工时间：2012
摄影师：©Nedyalko Nedyalkov (courtesy of the architect)

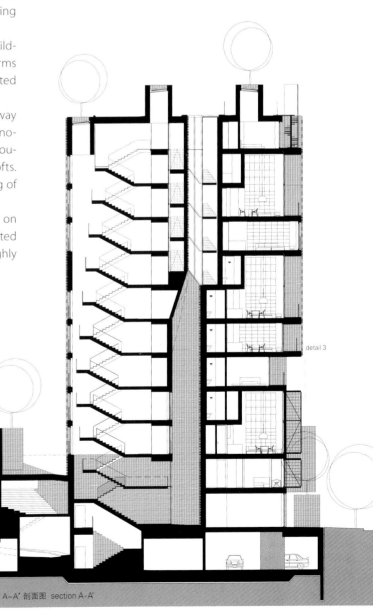

A-A' 剖面图 section A-A'

四层 fourth floor

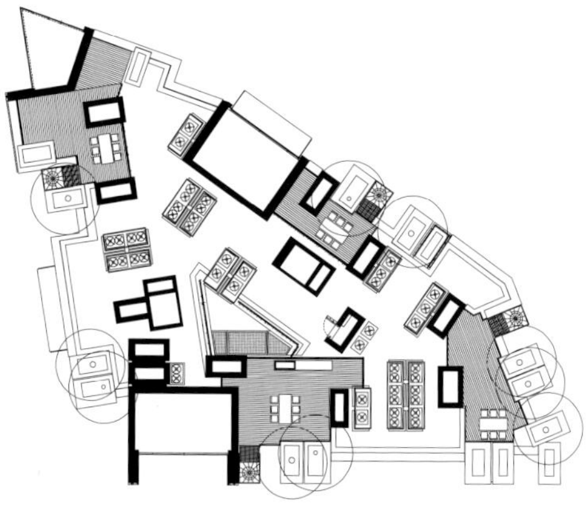

十三层 thirteenth floor

detail 2
三层 third floor

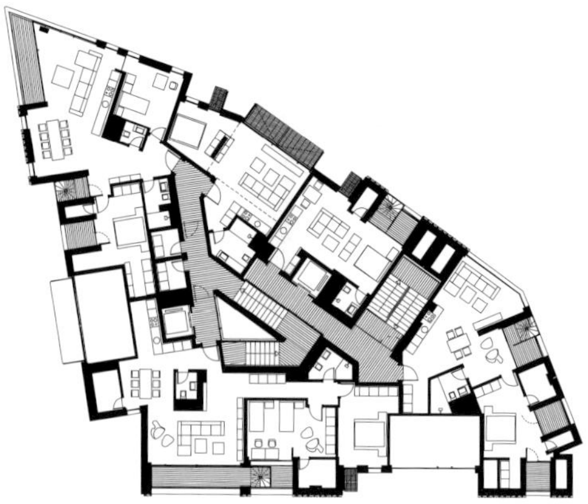

十二层 twelfth floor

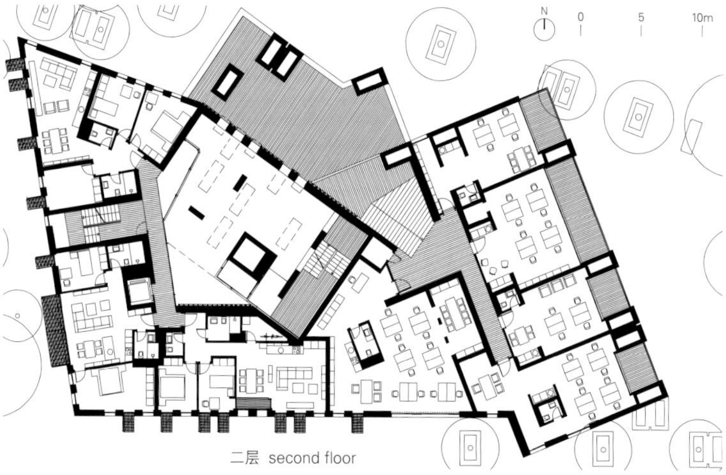

N 0 5 10m

二层 second floor

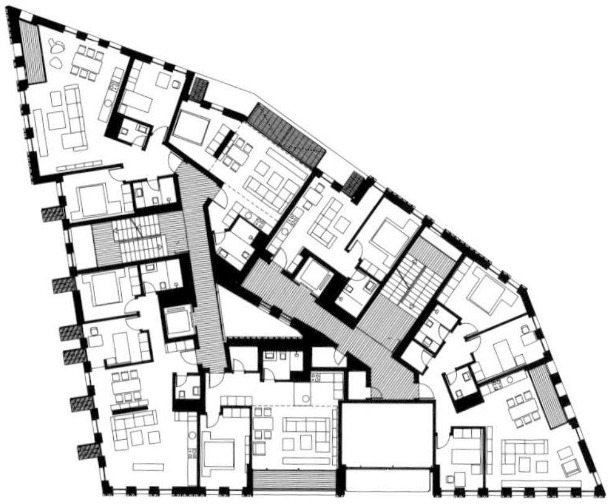

七层 seventh floor

detail 1
六层 sixth floor

一层 first floor

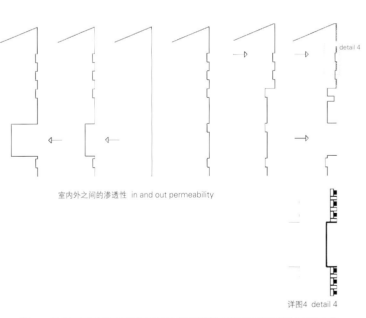

室内外之间的渗透性 in and out permeability

详图4 detail 4

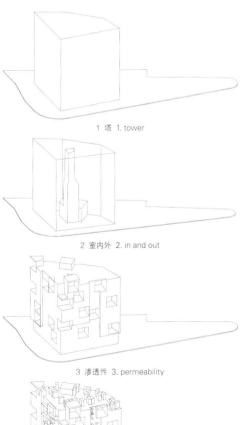

1 塔 1. tower

2 室内外 2. in and out

3 渗透性 3. permeability

4 规模 4. scale

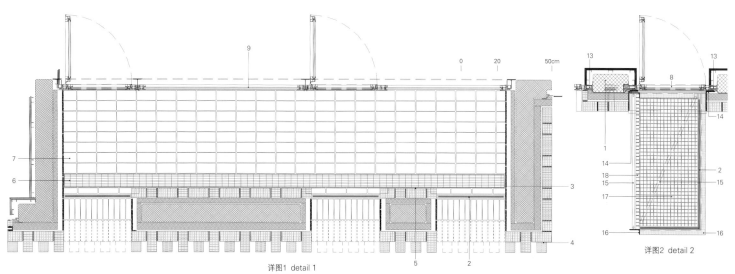

详图1 detail 1

详图2 detail 2

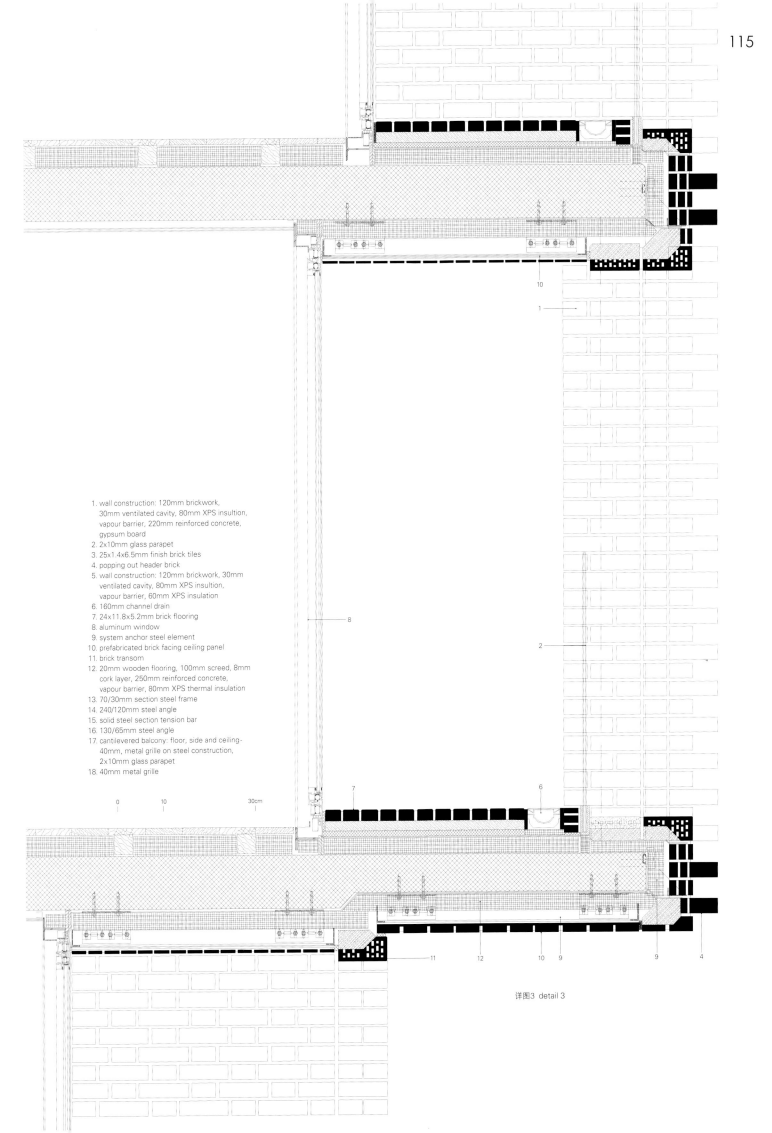

1. wall construction: 120mm brickwork, 30mm ventilated cavity, 80mm XPS insultion, vapour barrier, 220mm reinforced concrete, gypsum board
2. 2x10mm glass parapet
3. 25x1.4x6.5mm finish brick tiles
4. popping out header brick
5. wall construction: 120mm brickwork, 30mm ventilated cavity, 80mm XPS insultion, vapour barrier, 60mm XPS insulation
6. 160mm channel drain
7. 24x11.8x5.2mm brick flooring
8. aluminum window
9. system anchor steel element
10. prefabricated brick facing ceiling panel
11. brick transom
12. 20mm wooden flooring, 100mm screed, 8mm cork layer, 250mm reinforced concrete, vapour barrier, 80mm XPS thermal insulation
13. 70/30mm section steel frame
14. 240/120mm steel angle
15. solid steel section tension bar
16. 130/65mm steel angle
17. cantilevered balcony: floor, side and ceiling- 40mm, metal grille on steel construction, 2x10mm glass parapet
18. 40mm metal grille

详图3 detail 3

1. 10mm metal plate
2. 120/240mm steel angel
3. mounting elements
4. 30mm metal grille 40/40mm
5. metal grille 160/160 ø8mm
6. vertical metal bars ø8mm

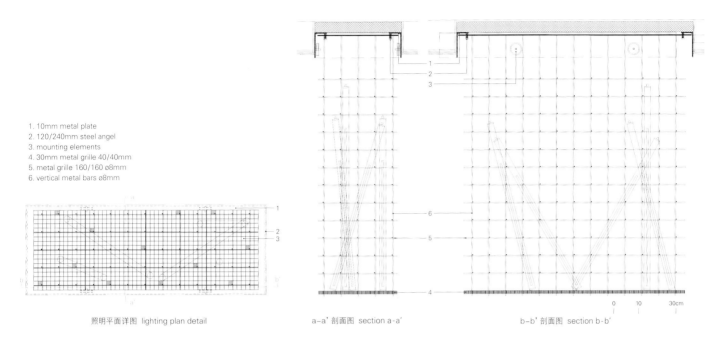

照明平面详图 lighting plan detail   a-a' 剖面图 section a-a'   b-b' 剖面图 section b-b'

服务于城市

人们越来越对大型偏远的权力中心没有信心，这一现象已经点燃了人们心中渴望的本地认同感的愿望，以此引发公民与本地机构和谐共存。出于这个原因，尤其是在非"古典"的时代，大型建筑物的特殊形象从以立体标志和权力象征为基础的传统性，转化为与城市肌理（社区能够更轻易地认清自我，即其本身的历史、记忆）相融合的理念。从这个意义上来说，当地机构的新场地不仅仅是通过发挥其政治功能，还通过提高城市空间的质量来服务于社会。如同Aldo Rossi判断的那样，历史告诉我们，为社会需求做出解释，是一座公共建筑应具有的功能，这些需求对其实际的纪念性特点起到了决定的作用，而这个特点通过具有进步性的功能累积以及与城市环境的融合而得以强化。通常，将本地机构的房屋连同地基加以升级，这一过程发生在与古老的公共建筑临近的区域，与各式辩证形式、水平以及融合手段所进行的评估息息相关。在大多数的情况下，当设计师在古建筑与新建筑之间建立一种连续性时，机构的代表性所扮演的角色是由古建筑的形式所遗留的产物来决定的，然而，对一个传统的大型建筑所产生的关于几个案例的研究仍然还在持续，尽管它们在现代语言中交织在一起。沿着这个方向，新市政厅建筑意图采用透明的建筑语言以及简单的石材表面，来故意避开华丽的形式。在这些建筑中，公共访问性以及小径的重要性将比任何形式的奢侈行为都重要。

A growing distrust for large, distant power centers has rekindled a desire for local identity and occasioned a reconciliation of citizens and local institutions. For this reason, especially in such non-"classical" times as these, the very image of monumentality has shifted from the tradition, built on solid icons and symbols of power, to the idea of an integration into the urban fabric in which the community can more easily recognize itself – its own history, its own memory. In this sense, the new sites of local institutions "serve the city" not only by exercising their bureaucratic functions, but by enhancing urban spatial quality. As Aldo Rossi has posited, history tells us that it is a public building's aptitude in interpreting community needs that determines its actual monumental character – a character enhanced by the progressive sedimentation of functions and integration into the urban context. Typically, the process of upgrading the premises of local institutions occurs in close proximity with ancient public buildings, with respect to which various dialectical forms and levels and means of integration can be assayed. In most cases the role of an institution's representativeness is left to what remains of the ancient forms, as designers establish a spatial continuity between the old and the new, but some examples of a search for a traditional monumentality persist, albeit articulated in a contemporary vocabulary. Along the way, in an evident intention to avoid any form of rhetoric, the architecture for new city halls tends to speak a language of transparencies and plain stony surfaces, in which public accessibility and the significance of paths assume more importance than any formal extravagance.

Serving the City

萨拉曼卡市政厅 / Carreño Sartori Arquitectos
韦克斯福德郡议会总部 / Robin Lee Architecture
毕尔巴鄂市政厅 / IMB Arquitectos
萨莫拉郡议会建筑 / G+F Arquitectos
莱希河畔兰茨贝格市政厅 / Bembé Dellinger Architekten
巴埃萨市政厅的修复 / Viar Estudio Arquitectura
Archidona的文化中心和市政厅
/ Ramón Fernández-Alonso Borrajo

服务于城市 / Aldo Vanini

Salamanca City Hall / Carreño Sartori Arquitectos
Wexford County Council Headquarters / Robin Lee Architecture
Bilbao City Hall / IMB Arquitectos
Zamora County Council / G+F Arquitectos
Landsberg am Lech Town Hall / Bembé Dellinger Architekten
Baeza Town Hall Rehabilitation / Viar Estudio Arquitectura
Cultural Center and the New City Hall of Archidona
/ Ramón Fernández-Alonso Borrajo

Serving the City / Aldo Vanini

美国政治学家Francis Fukuyama在他的文章《历史的终结与最后的人类》中说道，"国家是有目的性的政治产物，而人们在道德社区建立之前便存在了。也就是说，人类是怀有善与恶的信仰、了解神圣与亵渎的本质的团体，这些思想可能来自于对过去历史（但是现在主要作为一种传统）进行的深层挖掘。"[1]

中央集权机构公认的合法性急速地退化，而人们也发现了小型社区所具有的归属感。国家性问题与妥协或让步于权力无比强大的议员团体的需求所产生的不断扩大的复杂性，使公民从大众视角中不断地分离开来，而当地机构的利益也呈现出扩大的趋势。

公共建筑成为一个具体的、体现城市与机构关系的代表性建筑。因此，如果国家机构的宏伟的代表性建筑出现在国际著名且壮观的项目中，且远离普通的公民，如同公民感觉其远离这些项目所代表的机构一样，那么一种新趋势便会逐步地在本地建筑的规划中起决定作用。当隶属于小型团体的意愿超越国家公民身份的政治意义时，一座作为当地机构的建筑一定不是仅代表一个行政机构，在社区之内，它必须成为一处代表社区本身、文化、真实或者虚拟身份的地方。

在某种程度上，这一过程的产生方式与古老的、带有突出的纪念物，如大教室和城市塔楼的社区的鉴定过程相似。实际上，每座纪念物的特殊意义，尤其是先进的西方民主政治，相对于现在的标志性价值来说，更多的是与功能相关，尽管前者在许多当代公共建筑的例子中有所出现。沿着这个方向，如果古老的教堂或者城市塔楼用来主导城市环境，并且对权力以及人民之间的等级关系或者是金字塔关系加以模仿的话，那么当地权力的发源地便被寄予成为与城市和谐连接的一部分。

As American politologist Francis Fukuyama argued in his essay *The End of History and the Last Man*, "States are purposeful political creations, while peoples are pre-existing moral communities. That is, peoples are communities with common beliefs about good and evil, about the nature of the sacred and the profane, which may have arisen from a deliberate founding in the distant past but which now exist largely as a matter of tradition."[1]

The perceived legitimacy of centralized institutions has progressively declined, and people have discovered a sense of belonging to smaller communities. The expanding complexity of national problems and the need to resort to compromises and to concede to the demands of immensely powerful lobbies constantly increases the separation of the citizenship from a general vision, while interest in the local institutional reality deepens.

The architecture of public buildings has always been a physical representation of the city's relationship with institutions. Thus, if the architectural representation of great national institutions has emerged primarily in the spectacular projects of international superstars – as distant from ordinary citizens as the citizens feel distant from the institutions that those projects represent – a new trend has increasingly determined the architectural programs of local institutional buildings. Where the desire to belong to a smaller community overwhelms the political meaning of national citizenship, a building hosting a local institution must represent more than just a bureaucratic administration; it must become a place through which the community represents itself, its culture, its real or imaginary identity.

This process, in a way, can appear similar to the ancient identification of communities with such outstanding monuments as cathedrals or city towers. Actually, the very meaning of monument, in particular for advanced western democracies, is related more to function than to iconic value at present, although the latter is still present in many contemporary examples of public buildings. Along those lines, if the ancient cathedral or city tower stood up to dominate the urban context, mimicking the hierarchical or pyramidal relationship between the power and the people, now the home of local power is expected to integrate itself as a well-connected part of the city.

One result of the shift described above has been a loss of the typological identity that in the past functionally characterized and

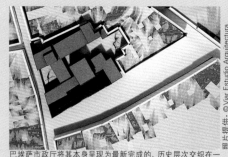

巴埃萨市政厅将其本身呈现为最新完成的、历史层次交织在一起的例子
Baeza Town Hall, presenting itself as the latest articulated example of historical stratification

Archidona的文化中心和市政厅,强化新旧建筑之间的过渡性
Cultural Center and the New City Hall of Archidona, enhancing the transition between the new and the existing

上述的转变导致的结果之一就是类型学特点有所缺失,这些特点都存在于过去那些功能各异、精心定义的公共建筑,如交通、市政厅、博物馆或是医院中。当这些项目在形式上致力于代表城市时,它们反之还会被加以构思和设计,来服务于城市,尽管"服务"这个词语不仅仅是与政府或者是行政功能相关联,还要以融合的方式,对城市的进化过程也起到辅助的作用。

形式和功能之间不是十分严格的关系既不是一种全新的也不是一种原始的现代运动理念。每个经历过历史和机构所产生的危机的时代都曾是没有在公共建筑中引用正确的类型学或者是语义分离(在特殊建筑的功能重利用的基础上)的结果。一个处于形式与功能之间且严格的语义依赖性具有社会和政治稳定性的时代的特点,这些我们便称之为"古典主义城市"。

现在,既然在民主管理进程中涉及到公民的空间容量,包括具体形式的,相比为社会提供一个标志来说更加重要,那么室内空间便要经常与室外,更重要的,与城市空间的宜居性保持密不可分的关系。一处适合居住的"空旷区"的重要性不亚于使人类住满的要求,而一条充满生机的小径比一座纪念物所带来的沉静更加重要。

城市建筑中形式和功能之间的模糊性,以及转变和分层过程中的历时性在Aldo Rossi的《科学自传》一书中得到了很好的阐述。

"……很明显,每一个物体都有一个其必须回应的功能,但是这个物体并不是在那个阶段结束,因为功能随着时间的变化而有所不同。这一直是我的一条相对科学的断言,并且我曾经将它从城市的历史和人类的生活中抽离:从宫殿、圆形剧场、修道院、住宅的改造或者是其环境的改造中抽离。……我曾经目睹过现在居住了许多家庭的宫殿,修道院被改造为学校,圆形剧场改造为足球场。这些改造的效果都非常好,建筑师和精明的管理员都不曾介入。……这一类型学的自由度一旦建立起来,作为一种问题形式,总是让我着迷。……它就像建筑中神圣的理念一样,一座塔既不是一个独立的权力形象,也不是一个宗教的象征。我想起了轻质房屋、葡萄牙Castello di Sintra的大型锥形烟囱,以及筒仓和烟囱。后者地处我们这个时代最美的建筑之

well-defined public buildings such as churches, city halls, museums or hospitals. Where those were formerly devoted to "representing" the city, they are now instead conceived and designed to "serve" the city, although the term "serve" is related not merely to bureaucratic or administrative functions, but to assisting, by integration, the evolution of the urban process.
A less rigid relationship between form and function is neither a totally new nor an original Modern Movement concept. Every period of historical and institutional crisis has seen the abandonment of precise typological references in its public architecture or semantic detachment based on the functional reuse of specific buildings. A strict semantic interdependence between form and function is characteristic of times of social and political stability, those we know as "classicity."
Since the capability to involve citizens in the democratic process of governance, including physically, is now more relevant than providing the community an icon, the internal space very often has an uninterrupted relation with the exterior and, more importantly, with the livability of the urban space. A livable "emptiness" is no less important than the imposition of "fullness," while a dynamic path is more important than a monumental stillness.
The ambiguity of form and function in the city's buildings and the diachronic process of transformation and stratification are well articulated by Aldo Rossi in his *Scientific Autobiography*:
"[...] it is evident that every object has a function to which it must respond, but the object does not end at that point because functions vary over time. This has always been a rather scientific assertion of mine, and I have extracted it from the history of the city and of human life: from the transformations of a palace, an amphitheater, a convent, a house, or of their various contexts. [...] I have seen old palaces now inhabited by many families, convents transformed into schools, amphitheaters transformed into football fields; and such transformations have always come about most effectively where neither an architect nor some shrewd administrator has intervened. [...] This freedom of typology, once established, has always fascinated me as a problem of form. [...] It is like the idea of sacredness in architecture: a tower is neither solely an image of power nor a religious symbol. I think of the lighthouse, the huge conical chimneys of the Castello di Sintra in Portugal, silos and smokestacks. The latter are among the most beautiful architecture of our time, but it would be untrue to say that they lack architectural models: this is another silly idea from modern or modernist criticism."[2]
The present is certainly not a "classical" time. Rather, it is a time of unexpected transformations, of radical rejection of the establishment, of a cultural inclination towards instability and ambiguity. Even amphitheaters and palaces are being completely

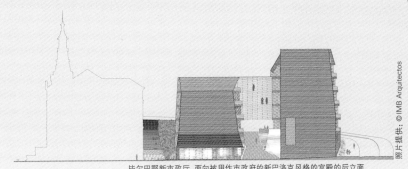

毕尔巴鄂新市政厅，面向被用作市政府的新巴洛克风格的宫殿的后立面
Bilbao New City Hall, facing the back facade of the Neobaroque Palace currently being used as Town Hall

间，但是不现实地说，它们缺乏建筑模型：从现代或者是现代主义的要求来看，这又是一个愚蠢的理念。[2]

现在当然不是一个"古典"的时代。相反，它是一个意想不到的、拒绝激进的社会体制的改造时代，也是文化倾向于不稳定和模糊的时代。尽管圆形剧场和宫殿被完全地改造——出于在动荡年代的安全因素，古老城市出现了一种实际的且普遍的内爆症状——而这些对城市空间进行精心利用的现代意识是反对城郊地区的离心式浪费的，社区通过重新思考它们的功能，或者是使其转变为全新的环境和公共建筑复合体，来拯救老建筑，或者是其遗迹。

由于这些原因，"服务于城市"的规划成为一个有趣的挑战，它在表现与服务之间、大型建筑与综合性建筑之间、城市辨认性与室外/室内的连续性之间展开。在这一过程中，以某种理念为依据，时间和记忆与空间融合在一起，而20世纪的物理学通过这一理念，已经彻底地改变了我们关于现实的观念。

Viar Estudio建筑师事务所在其位于西班牙的巴埃萨市政厅的修复项目中，完美地诠释了这些理念。该项目没有作为一个竣工的、固有的建筑，而是将其本身呈现为最新完成的、历史层次交织在一起的例子，并且为将来城市空间"持续性"的进化做出贡献。建筑师承认受到了Henri Bergson的"durée"理念的启发：时间作为一个自身正在增长的时刻，与其他时刻重叠在一起。这个项目被构思为旧城市的一个完成部分，并不仅仅是对遗留部分进行修复，成为旧市政厅的一个华丽的部分，而是要成为一个逻辑序列的产物，包括古老的立面，通过移除不适合的多余部分，在巨大的建筑内加以突出。

建筑成为一个容纳办公室、公用区和顾客服务区的集装箱，其所连接的公共空间没有解决它与街道的连续性。简而言之，时间与记忆、功能与丰富性、可居住的体量——这三个逻辑方面——都是彼此亲密联系的。古老的立面通过一个巴洛克式楼梯被投影在行政办公室中，楼梯是嵌入结构的中心节点，而旧的拱顶和拱门与现代的、简约的白色表面融合在一起，对更适宜的机构空间的标志性价值起到保护的作用。石质地面从公共街道开始向内部小院延伸，使小院不仅仅是一座庭院，还是

transformed – one physical symptom of the general implosion of ancient cities for reasons of safety in unsafe times – the present consciousness of the careful use of urban space has arisen against the centrifugal waste of suburban territories, as communities rescue old buildings, or what remains of them, by rethinking their function, metabolizing them into new contexts and public compounds.

For these reasons, the program for "serving the city" becomes an intriguing challenge that unfolds on the ground between representation and service, monumentality and integration, urban recognizability and external/internal continuity. In such a process, time and memory blend with space in accordance with certain concepts by which twentieth-century physics has definitively revolutionized our notions of reality.

In his Baeza Town Hall Rehabilitation, in Spain, Viar Estudio Arquitectura perfectly interprets all these concepts. Rather than as an accomplished, immanent building, the project presents itself as the latest articulated example of historical stratification and as a further contribution to the evolutionary "duration" of the urban space. The architect admits to being inspired by Henri Bergson's "durée": time as a moment that grows in itself, overlapping other moments. Conceived as the completion of part of the old city, the project is not merely a rehabilitation of the remaining and a rhetorical fragment of the old Town Hall, but the result of a logical sequence that includes the ancient facade, enhanced in its monumentality by the cleaning up of every improper addition.

It becomes the container of offices, the common and customer service area, and the public space to which it is connected without resolving its continuity with the street. In a few words, time and memory, function and fullness, the livable void – the three logical aspects – are here intimately interconnected. The historical facade is projected toward the administrative offices through a baroque staircase that is the central node of the intervention, while the old vaults and arches blend with the contemporary, minimalistic white surfaces, preserving the iconic value of the more properly institutional space. The stone floor extends itself from the public street toward the internal court, which is more of a continuation of the street than a mere courtyard. The interior geometry suggests an articulation of a public open space and, like an old town street, narrows and widens in directions that are not always parallel.

Bembé Dellinger Architekten work less subtly than de Viar Estudio Arquitectura, although with a similar approach, in their addition to the Town Hall of Landsberg am Lech in Germany. The new extension imposes its presence as a great, dark volume, clad in a very material copper net that nevertheless allows the mutual

韦克斯福德郡议会总部,不仅仅要避免建造一个带有标志的纪念碑的诱惑

Wexford County Council Headquarters, not only avoiding the temptation to create a monument with an iconic presence

街道的延伸。室内的几何外形显示了公共开放空间的连接方式,如同古老的城镇街道一样,在各个方向(非平行)进行紧缩和拓宽。

在位于德国的莱希河畔兰茨贝格的市政厅增建项目中,Bembé Dellinger建筑事务所尽管采用了一个类似的方法,但是其工作量相对于de Viar Estudio建筑师事务所来说,还是少了些许。新扩建的部分呈现为一个宏伟的黑色体量,覆以特殊的金属网,不过却允许室内外的光线进入。虽然一层尽量保持空旷,以更好地通过老建筑来连接城市空间,新扩建的部分在现存的环境中处于完全封闭的状态,没有将其本身呈现为一座全新的地标,而是将这一功能留给了市政厅的立面。

一个更深层且更复杂的融合以及"城市服务"层次出现在萨莫拉省的Diputación的新郡议会建筑中。Pilar Peña Tarancón and María Antonia Fernández Nieto将这座行政建筑构思为萨莫拉Viato广场大型公共空间的一个元素。项目意图保留着其微妙的标志性,使其相对于建筑的内在形式来说,与广场特定的形态特征的联系更加亲密。其普通的表面形成了具有高度象征性的广场的第四个侧面,并且广泛地使用本地石材,使人们回忆起城墙,建筑没有隐藏其内部的钢结构,来表明其同时代性。体量间的相互连接留出了一处大型面向广场的上空空间,这处空间类似于阳台,为人们提供了望向公共空间或者从室内庭院望去的视野。建筑师极其关注自然照明,来突出建筑的"公共"特点,以及其在广场活力中具有的归属感。

IMB建筑事务所采用不同的规模,将融合的水平层次转移到巨型建筑物的地面中来。这座毕尔巴鄂市政厅的总部面对着古老的折中风格的宫殿(容纳市政府)的后立面,采用将其本身置在一个完全不同维度的方法,在城市景观中扮演着全新的角色。以牺牲老建筑为代价来吸引人们眼球的这一目的十分明显,不仅仅是从增建部分的维度来看,而且还从运用了动态的、非正交的体量和透明表面,与邻近建筑的古典式静态形成对比的方面也可以看出。尽管毕尔巴鄂市政厅拥有现代的外形,但是它仍然属于机构建筑将自身展示为社区的化身这一传统。总之,相比较成为城市肌理的一部分,它更是一座教堂。鉴于它

passage of external and internal light. Although the ground floor is kept as empty as possible for easier connection to the urban space through the old building, the new extension is completely enclosed in the existing context, not presenting itself as a new landmark, but leaving this function to the Rathaus facade.

A deeper and more complex level of integration and "urban service" emerges in the new county council for Diputación Provincial de Zamora. Pilar Peña Tarancón and María Antonia Fernández Nieto have conceived the administrative building as an element of the large public space of Zamora's Viato Square. The project maintains a subtle iconic intention, more related to the specific morphological character of the square than to the building's intrinsic form. Its plain surface completes the fourth side of a highly symbolic square, adapting itself to generalized use of the local stone that also recalls the city walls, even while evincing its contemporaneity by not hiding its internal steel structure. The articulation of the volumes leaves a large void facing the square that, balcony-like, offers a view of the public space or from the internal courtyard. The extreme attention to natural illumination reinforces the building's "public" character and its sense of belonging in the dynamics of the square.

Operating on a different scale, IMB Arquitectos shift the level of integration onto the ground of monumentality. Facing the back facade of the old eclectic palace that hosts the Town Hall, the headquarters building of Bilbao City Hall plays a new role in the urban scene by positioning itself in a different dimensional scale. The intention to attract attention at the expense of the old building is clear, not only in terms of the addition's dimensions, but also by means of its dynamic, non-orthogonal volumes and its transparent surfaces, in contrast with the classical staticity of the adjacent building. Despite its contemporary shape, the Bilbao City Hall belongs to a tradition of institutional buildings presenting themselves as the incarnation of the community. In a word, it is more a "cathedral" than a part of the urban fabric. Given that it is no longer possible to employ traditional signs of institutional power, the authors rely on a current lexicon of monumentality based on the imposing posture of the volumes and their non-orthogonal dynamics.

Greater attention to the preexisting context is evident in the rehabilitation of the Old School Square in Archidona, Spain, by Fernandez Alonso, which evokes an intention that is not merely formal, but also functional and programmatic. The project's declared

萨莫拉郡议会建筑，呈现出相当友好的风格，即一个优雅的公寓社区，面向社区开放
Salamanca City Hall, offering the friendly patterns of a elegant condominum opening itself to the community

不再有可能采用传统机构的权力标志，建筑师便以强加体量的姿态以及其正交性的动态为基础，依赖于巨型建筑现拥有的所有要素。

在西班牙Archidona的旧学校广场的修复项目中，建筑师将更多地关注之前存在的环境这一意图非常明显，该项目由Fernandez Alonso设计，唤起了形式、功能以及程序方面的意图。该项目宣称的意图是强化新旧建筑之间的过渡性，如同de Viar Fraile的项目一样，它是旧建筑和现代服务功能中最具有代表性的功能。新与旧完全融为一体，与壮观的Ochavada广场相融合。该建筑综合体的室内小径成为广场的八角空间的一个离心式投影。

通过爱尔兰皇家建筑机构发起的开放式设计竞赛来设计新总部，韦克斯福德郡议会总部一定会采用一个"服务于社会"的规划，不仅仅是要避免建造一个带有标志和传统意义的纪念碑的诱惑，同时还要拒绝明星建筑师的标记所具有的可视性。项目的"公共价值"被移交给提议本身。Robin Lee将他的项目作为一个功能性的综合公共服务园区，它能与社区建立关系。Lee的作品中没有任何夸张华丽的暗示，而是利用透明性、反射以及石材的沉重性来处理公共区和服务区的序列问题。

人们可能会注意到由Carreño Sartori建筑事务所在智利的萨拉曼卡的采矿小镇Choapa山谷中设计的新市政厅项目中，建筑师采用了更进一步的方法来把项目与现存的环境相融合。在这个案例中，城市形态成为海拔较低的建筑环境（与常规的正交城市网格相匹配）内一个相互尊重、非壮观的嵌入结构中的关键。建筑将自身呈现为网格的一个按比例扩大的元素，将其自身限制在同样严格的正交外形中。此外，它没有任何一座机构建筑常见的标志，而是呈现出相当友好的风格，即一个优雅的公寓社区，面向社区开放，设有一个斜坡体系，引导人们从地面层走向顶层，同时连接着各式各样的市政服务区。

在上述的大多数例子中，与建筑"肌肉"的夸张式展现不同的是，公共建筑反对夸张式理念，尤其是当地机构被视为临近社区的直接财产，而非一些偏僻的、至高权力的体现，这种上升的趋势是十分明显的。

purpose is to enhance the transition between the new and the existing, placing, as de Viar Fraile's project did, the most representative function in the ancient section and the service functions in the contemporary. Old and new blend in absolute continuity, in a integration with the manificent Plaza Ochavada. The complex's internal paths act as a centrifugal projection of the square's octagonal space.

Opting to design its new headquarters through an open design competition run by the Royal Institute of the Architects of Ireland, the Wexford County Council Headquarters certainly adopted a program to "serve the city," not just avoiding the temptation to create a monument with an iconic presence and a traditional meaning, but also rejecting the visibility of an archistar signature. The "public value" of the project was handed over to the proposal itself. Robin Lee integrated his project as a functionally complex public service campus which establishes relationships with its community. Avoiding any hint of extravaganza, Lee plays with transparencies, reflections and stony heaviness to manage the sequence of public and working areas.

One may observe a further approach to integration into the existing context in the new City Hall of the mining town of Salamanca, Choapa Valley, Chile, by Carreño Sartori Arquitectos. In this case, urban morphology is the key to a respectful, anti-spectacular insertion in the context of low buildings aligned to a regular, orthogonal grid. The building presents itself as a scaled-up component of that grid, constraining itself to the same rigorous orthogonal geometry. Furthermore, it offers none of the usual signs of an institutional building, but rather the friendly patterns of an elegant condominium opening itself to the community by means of a ramp system leading from the ground to the top floor and connecting the various municipal services.

In the majority of the above examples, as opposed to a megalomaniacal exhibition of architectural muscle, there is an evident increasing tendency toward an anti-rhetorical conception of public buildings and, in particular, toward the premises of local institutions being seen as proximal, direct properties of the community, rather than as embodiments of some distant, superordinate power. Aldo Vanini

1. Francis Fukuyama, *The End of History and the Last Man*, The Free Press, New York, 1992.
2. Aldo Rossi, *A Scientific Autobiography*, The MIT Press, Cambridge MA, 1981.

萨拉曼卡市政厅
Serving the City
Carreño Sartori Arquitectos

服务于城市

萨拉曼卡市位于内陆山谷中,常驻居民为12 000位,处于乔阿帕河盆地的起源处,一处富有矿产的乡村区域。其背景景观为荒芜的垂直斜坡,它包含设有一条长长的且狭窄的农业带的山谷。这个项目在现今较低的建成环境中形成了一个全新的规模。这一新型高度不但被认为是一座引人注目的塔楼,而且还是一座关注其与室内空间和周围山谷之间的特殊连接的五层建筑。

场地临近城市的主要广场,即萨拉曼卡的de Armas de广场。场地给人的第一直觉是这次建筑之旅要从一条扩大的公共铺路开始,将建筑作为城市环境中的一部分。它可以被理解为一处拓展的公共空间:一条从街道水平延伸至建筑顶部平台的斜坡,穿过室内空间,面向景观开放,将各种市政设施聚集在一起。

建筑师理解一座市议会的所需并不仅仅是一座简单的办公建筑,那么这个项目便会成为一种为人们提供多种用途的"大房子"类型。

在古老的希腊城市中,公共活动经常发生在市场中,而在殖民城市中,从同样的地中海文化中来看,则发生在露天的市议会建筑中。如今,这些机构的需求变得些许不同,但是建筑师一直谨记它是一处"会见场所",并将其作为整个项目最基本的理念。

今天的政府体系需要足够的功能空间来聚集一系列的办公室和设施,它们可以被归纳为工作站、会议室以及行政办公室。同时,这座建筑每天要欢迎大量的人们,即随时到访的市民,这将会使建筑的规划更加复杂,其公共的性质更要得以巩固。

这两个新世界将在一处开放的建筑内交汇,并且通过一个斜坡连接在一起。

建筑的主要结构为一个建在场地上的带有抗震屋顶的混凝土结构,而所有其他的结构构件,如混凝土嵌板、斜坡以及窗户,都是预制的,并且利用灵活的连接(消减地震运动)安装在主结构之上。

楼层平面布局较为自由,被分成两个半错层,由斜坡所处的室内上空空间分隔开。两个不同的独立楼梯和电梯在室内提供了有效的交通流线。

在地面至五层的上升过程中,斜坡穿过更加开放的功能区和社区,到达需要更多亲密性和独立性的功能区。

空间沿着斜坡采用纵向分隔方式,相反,功能则采用横向分隔。因此穿过不同的层次则代表着一个从接待处到每个执行部门的层次序列。

一些特殊的考虑也被建筑师加以关注,如舒适性、朝向、用途、照明以及通风等。例如,窗户和庇荫系统的设计根据太阳路径也有所不同,因此形成了较为独特的室内空间,人们可在此享受自然光。

Salamanca City Hall

The city of Salamanca (12,000 inhabitants) is located in an inner valley, at the origin of the Choapa River Basin, a minning and rural area. A background landscape of vertical arid slopes contains a valley of long and narrow agricultural strip. This project introduces a new scale in the current low height built environment. This new height is not conceived as an outstanding tower, but rather as a five floors building caring about a special connection with its intern spaces and the surrounding valley.

The site is adjacent to the city main square, Plaza de Armas de Salamanca. The first intuition was to initiate the building tour from an enlarged public pavement, thinking the building as part of the urban context. It can be understood as an extended public space: a ramp runs from the street level up to a terrace on the top of the building, through an interior space, opened to the landscape, that brings together the various municipal facilities.

Through the understanding of what a town council needs, more than a simple office building, the project could become a type of "big house", that gives space to diversity of uses and people.

In the Ancient Greek Cities, public life used to take place in the agora; in the colonial cities, from the same mediterranean culture, in the outdoor town council. Nowadays, the institutions' needs are slightly different, but the architects kept in mind the idea of a "meeting place" as the fundamental concept of the project.

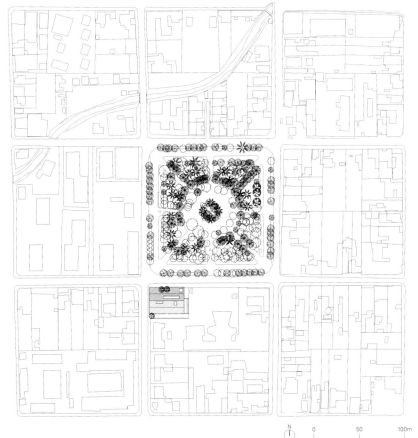

北立面 north elevaition   0 5 10m   西立面 west elevation

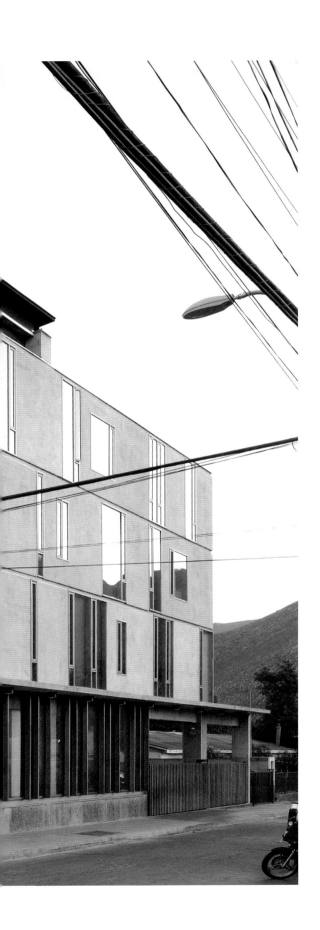

Today's government systems require spaces functional enough to gather a series of offices and facilities that could come down to workstations, meeting rooms, and executive offices. Meanwhile, the building has to daily welcom a large number of people, that is, random citizens, which bring complexity to the building's program and consolidate its public nature.

Those two worlds would meet in an open space building, articulated around the ramp.

The building's main structure is a seismic-proof concrete structure erected on site, while all other construction elements – like concrete panels, ramps and windows – are prefabricated and assembled with the main structure with flexible connections that dissipate seismic movements.

The free plans are divided in two half floor, separated by the inner void where is located the ramp. Two different independant staircase and elevator provide an efficient circulation inside the building.

In this rise from the urban ground to the fifth floor, the ramp goes through programs more opened to the public and the community to programs requiring more intimacy and independency.

The space division is longitudinal along the ramp and, on the contrary, the program division is transversal, thus crossing different layers representing a hierarchical order going from the reception desk to the executive of each department.

Special considerations have been taken regarding comfort, orientation and uses, natural lighting and ventilation. I.e, the different designs of the windows and shading systems according to the path of the sun, thus creating singular inner spaces, so people can enjoy the natural lighting.

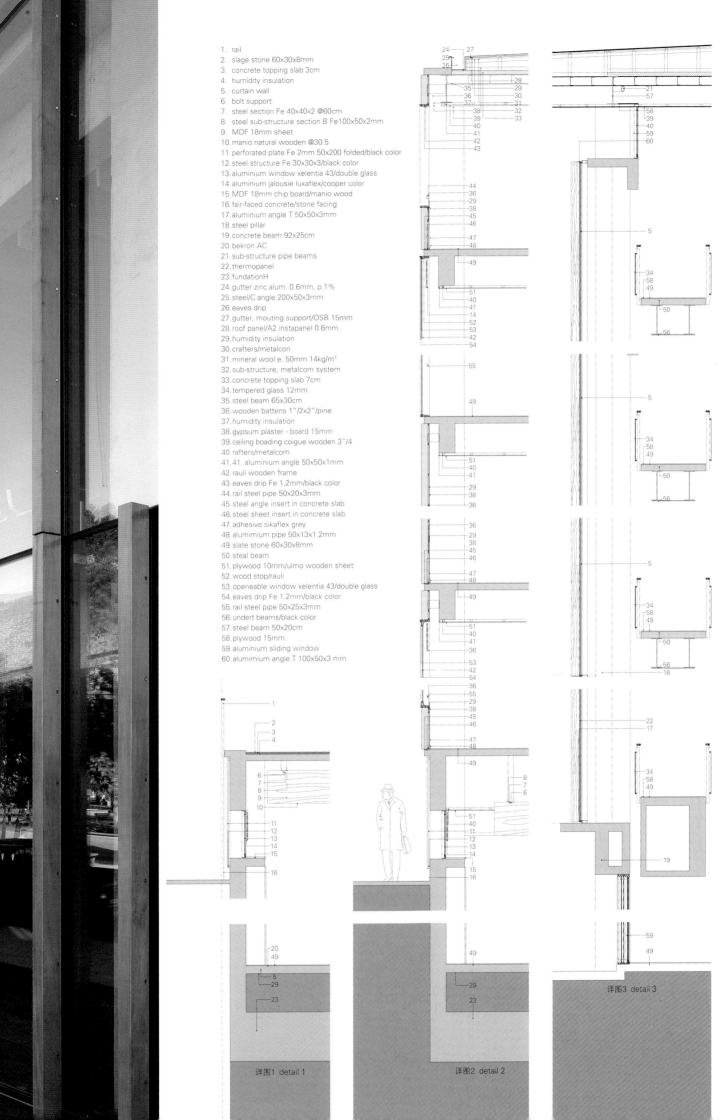

1. rail
2. slage stone 60x30x8mm
3. concrete topping slab 3cm
4. humidity insulation
5. curtain wall
6. bolt support
7. steel section Fe 40x40x2 @60cm
8. steel sub-structure section B Fe100x50x2mm
9. MDF 18mm sheet
10. manio natural wooden @30.5
11. perforated plate Fe 2mm 50x200 folded/black color
12. steel structure Fe 30x30x3/black color
13. aluminium window xelentia 43/double glass
14. aluminium jalousie luxaflex/cooper color
15. MDF 18mm chip board/manio wood
16. fair-faced concrete/stone facing
17. aluminium angle T 50x50x3mm
18. steel pillar
19. concrete beam 92x25cm
20. bekron AC
21. sub-structure pipe beams
22. thermopanel
23. fundationH
24. gutter zinc alum. 0.6mm, p.1%
25. steel/C angle 200x50x3mm
26. eaves drip
27. gutter, mouting support/OSB 15mm
28. roof panel/A2 instapanel 0.6mm
29. humidity insulation
30. crafters/metalcon
31. mineral wool e. 50mm 14kg/m²
32. sub-structure, metalcom system
33. concrete topping slab 7cm
34. tempered glass 12mm
35. steel beam 65x30cm
36. wooden battens 1"/2x2"/pine
37. humidity insulation
38. gypsum plaster - board 15mm
39. ceiling boading coigue wooden 3"/4
40. rafters/metalcom
41. 41. aluminium angle 50x50x1mm
42. raulí wooden frame
43. eaves drip Fe 1.2mm/black color
44. rail steel pipe 50x20x3mm
45. steel angle insert in concrete slab
46. steel sheet insert in concrete slab
47. adhesive sikaflex grey
48. alumimium pipe 50x13x1.2mm
49. slate stone 60x30x8mm
50. steal beam
51. plywood 10mm/ulmo wooden sheet
52. wood stop/raulı
53. openeable window xelentia 43/double glass
54. eaves drip Fe 1.2mm/black color
55. rail steel pipe 50x25x3mm
56. undert beams/black color
57. steel beam 50x20cm
58. plywood 15mm
59. aluminium sliding window
60. alumimium angle T 100x50x3 mm

详图1 detail 1

详图2 detail 2

详图3 detail 3

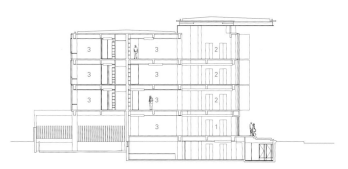

1 大厅 2 等候室 3 办公室
1. hall 2. waiting room 3. office
A-A' 剖面图 section A-A'

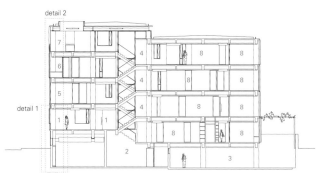

1 餐馆 2 礼堂 3 停车场 4 等候室
5 当地的治安法庭 6 市长办公室 7 平台 8 办公室
1. restaurant 2. auditorium hall 3. parking 4. waiting room
5. local police court 6. mayor 7. terrace 8. office
B-B' 剖面图 section B-B'

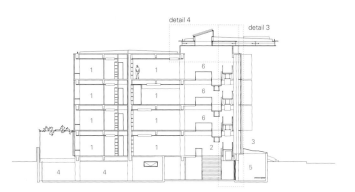

1 办公室 2 大厅 3 礼堂的照明庭院 4 停车场 5 礼堂 6 等候室
1. office 2. hall 3. auditorium light yard 4. parking 5. auditorium 6. waiting room
C-C' 剖面图 section C-C'

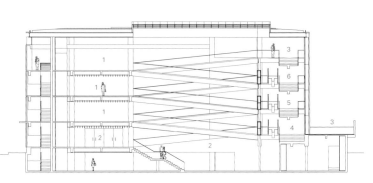

1 等候室 2 大厅 3 平台 4 餐馆 5 当地的治安法庭 6 市长办公室
1. waiting room 2. hall 3. terrace 4. restaurant 5. local police court 6. mayor office
D-D' 剖面图 section D-D'

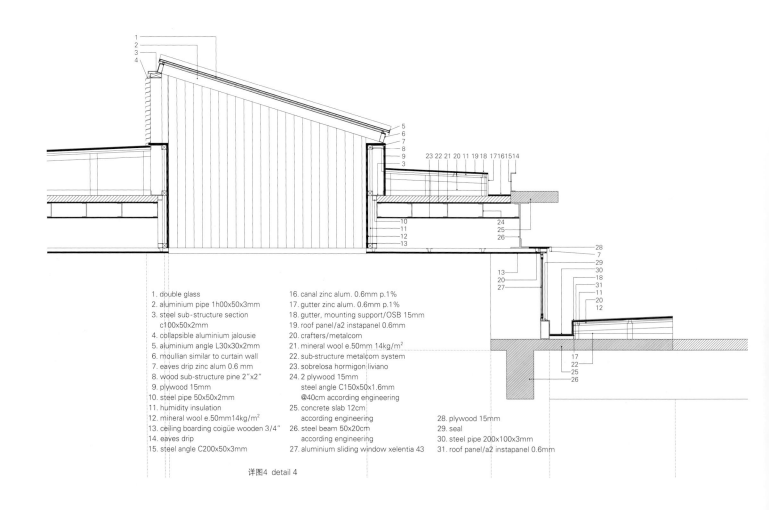

1. double glass
2. aluminium pipe 1h00x50x3mm
3. steel sub-structure section c100x50x2mm
4. collapsible aluminium jalousie
5. aluminium angle L30x30x2mm
6. moullian similar to curtain wall
7. eaves drip zinc alum 0.6 mm
8. wood sub-structure pine 2"x2"
9. plywood 15mm
10. steel pipe 50x50x2mm
11. humidity insulation
12. mineral wool e.50mm14kg/m²
13. ceiling boarding coigüe wooden 3/4"
14. eaves drip
15. steel angle C200x50x3mm
16. canal zinc alum. 0.6mm p.1%
17. gutter zinc alum. 0.6mm p.1%
18. gutter, mounting support/OSB 15mm
19. roof panel/a2 instapanel 0.6mm
20. crafters/metalcom
21. mineral wool e.50mm 14kg/m²
22. sub-structure metalcom system
23. sobrelosa hormigon liviano
24. 2 plywood 15mm
    steel angle C150x50x1.6mm
    @40cm according engineering
25. concrete slab 12cm according engineering
26. steel beam 50x20cm according engineering
27. aluminium sliding window xelentia 43
28. plywood 15mm
29. seal
30. steel pipe 200x100x3mm
31. roof panel/a2 instapanel 0.6mm

详图4 detail 4

项目名称：Salamanca City Hall
地点：Salamanca, Región de Coquimbo, Chile
建筑师：Mario Carreño Zunino, Piera Sartori del Campo
合作商：Pamela Jarpa Rosa
结构工程师：SyS. Mauricio Sarrazin A.
电气工程师：ICG S.A.
承包商：Constructora INCOBAL
甲方：Municipalidad de Salamanca
用地面积：1,280m²
总建筑面积：4,400m²
有效楼层面积：980m²
材料：reinforced concrete, reinforced concrete prefabricated pannels, double glass windows, Raulí wood, Coigüe wood
设计时间：2008
竣工时间：2010
摄影师：©Cristóbal Palma(courtesy of the architect)

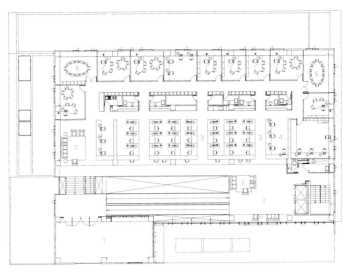

1 平台 2 等候室 3 办公室 4 当地的治安法庭 5 会议室
1. terrace 2. waiting room 3. office 4. local police court 5. meeting room
四层 fourth floor

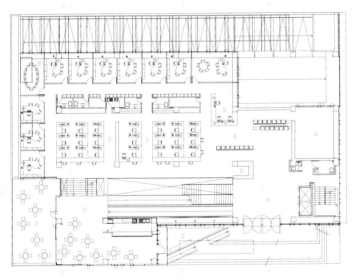

1 平台 2 餐馆 3 礼堂的照明庭院 4 大厅
5 接待处 6 等候室 7 办公室 8 会议室
1. terrace 2. restaurant 3. auditorium light yard 4. hall
5. reception 6. waiting room 7. office 8. meeting room
一层 first floor

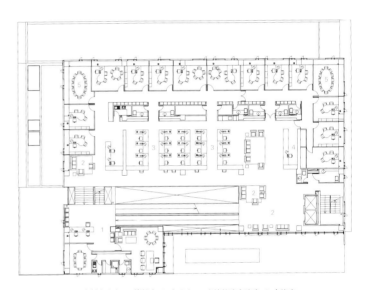

1 市长办公室 2 等候室 3 办公室 4 当地的治安法庭 5 会议室
1. mayor office 2. waiting room 3. office 4. local police court 5. meeting room
三层 third floor

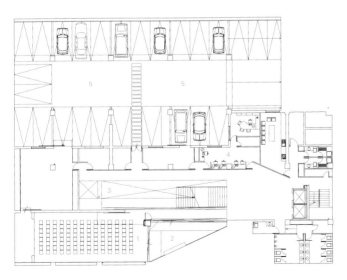

1 礼堂 2 礼堂的照明庭院 3 大厅 4 收银台 5 停车场
1. auditorium 2. auditorium light yard 3. hall 4. cash desks 5. parking
地下一层 first floor below ground

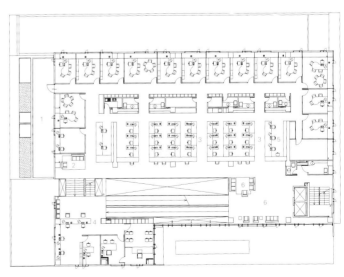

1 平台 2 等候室 3 办公室 4 当地的治安法庭 5 秘书处 6 等候室
1. terrace 2. waiting room 3. office 4. local police court 5. secretary 6. waiting room
二层 second floor

韦克斯福德郡议会总部
Robin Lee Architecture

Serving the City

服务于城市

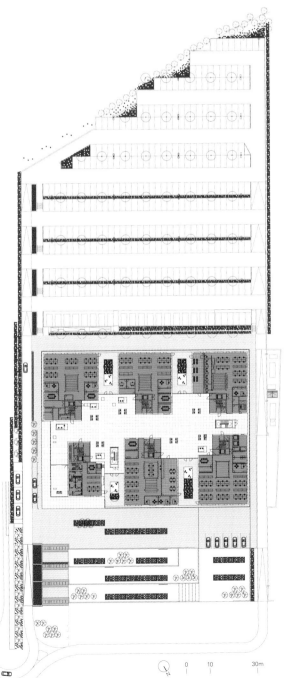

这座新的市政中心将韦克斯福德郡议会的服务与部门结合在一起,直到现在,这个郡议会中心都被分开设置,位于小镇的中心内。建筑赋予了议会所做的集体努力一种认同,将其作为一个统一的机构,同时赋予每个独立的部门及其特殊活动一种个性表达。

建筑布局为一系列六个分开的体块:每个体块容纳重要的服务设施和独立部门。这些体块聚集在一起,围绕在一处大型中心空间的周围。市民论坛是一处举办市政典礼、演讲以及社会聚集活动的空间,成开放状态,人们都可以进入其中,同时它也是所有市政设施的门厅。

郁郁葱葱的庭院栽满了当地的植物,且还有注满了净水的清澈水池,它们将体块分隔开来。这些庭院为建筑更深的部分带来阳光,同时将室内与周围景观连接起来。这些空间结合起来,将社交空间安排在建筑的中心,且允许公共领域入侵到建筑的整个一层中。

墙壁和地板被覆以爱尔兰的蓝色石灰石,创造了一个雕塑般的室内体量,其氛围安静且雅致。

建筑坐落在韦克斯福德镇外围边缘的一个斜坡场地上,地处爱尔兰的东南部,具有望向Slaney Estuary河以及Blackstairs山脉的良好视野。室外的玻璃层将体块包裹起来,成为双层立面的外层表皮。它在暴露的场地中发挥着保护的作用,同时也通过对周围进出空气的控制来调节室内的温度。在夏天,建筑内非常凉爽,而在冬季,它则发挥着保温层的作用。

玻璃立面进行统一处理,低铁玻璃黏合在阳极氧化铝框上,以形成一个透明的外围护结构,赋予建筑一种独特性和一致性,且其规模也符合其市政的身份。

阶梯式平台和景观花园改善了斜坡式的场地,抬升了建筑,并且为建筑创造了一处带有入口的景观和市政背景。

建筑为现浇混凝土结构,重混凝土制成的拱腹暴露在外,且贯穿公共和办公空间,以从热质量高的混凝土性能中获益。在公共广场中,暴露的混凝土拱腹与主要的石板墙体和地面相结合,形成一个无菌的环境。这一措施对防火策略起到了支持的作用,因为广场还发挥着建筑体块内独立部门的主要防火逃生路线的作用。这处空间内暴露的拱腹经过了精挑细选,赋予了空间体验一种沉重感和庄重性。同样重要的是,它还要创造一种强烈的表现感,以和功能保持一致,发挥着重要且尊贵的市政空间的作用。

Wexford County Council Headquarters

The new civic headquarters brings together the services and departments of Wexford County Council that, until now, has been housed separately within the center of the town. The building gives identity to the collective endeavour of the council as a unified organization while giving individual expression to the separate departments and their unique activities.

The accommodation is laid out as a series of six discrete blocks; each block houses key services and individual departments. The blocks are gathered around a large central space, a "civic forum", which is open and fully accessible functioning as a space for civic ceremonies, presentations and social gatherings as well as a foyer to all of the council facilities.

Lush courtyards filled with native planting and serene pools of still water separate the blocks. These courtyards bring light into the deeper portions of building and connect the interiors with the surrounding landscape. The spaces combine to place social space at the heart of the building and allow the public realm to pervade the whole building at ground floor.

Walls and floors are clad in Irish blue limestone, creating a sculpted interior volume with a calm, refined atmosphere.

The building sits on a sloping site on the outer fringes of Wexford Town, in Southeastern Ireland, with fine views to the River Slaney Estuary and the Blackstairs Mountains. An outer layer of glass

1. glass louvre blade
2. continuous timber fillet
3. louvre motor
4. WBP ply
5. roof membrane
6. continuous timber fillet
7. clear silicone seal
8. roof paving
9. 2mm aluminum powder coated pressing
10. angle cleat
11. continuous support angle powder coated
12. hot dipped galvanized steel angle wind ties
13. 40mm limestone cladding
14. EPDM bonded to concrete
15. glass vent panel sunstop silber 20
16. 8mm toughened low-E glass
17. anodised mullions and transom pale umber
18. blind
19. raised access floor
20. trench heater
21. oak veneer window board
22. plasterboard lining
23. powder coated steel walkway grating
24. powder coated steel bracket
25. ipasol fixed glazing unit
26. 2mm powerded coated aluminium flashing

详图1 detail 1

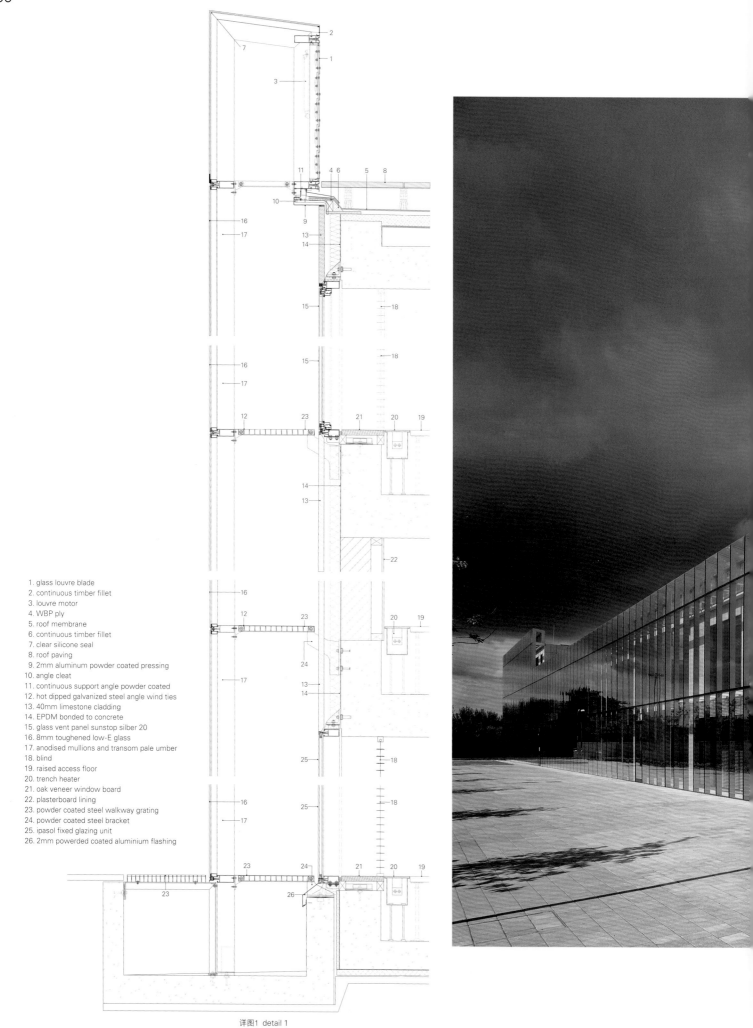

wraps around the blocks and acts as the outer skin of a double facade. This provides protection on an exposed site but also regulates the interior temperature through the control of air around the building; cooling the building in the summer and creating an insulating layer during the winter.

The glass facade is treated uniformly with structurally bonded low iron glass on anodized aluminum mullions to create a sheer envelope that gives the building a single, coherent identity and scale appropriate to its civic status.

This approach allows the interior spaces to find appropriate form, varied character and atmosphere while the outer form deals independently with scale and identity.

Stepped terraces and landscaped gardens ameliorate the sloped site, elevating the building and also creating an entrance landscape and civic setting for the building.

The central space is informed by the dense, rich townscape of Wexford town and the distinctively flat, empty, landscape of the River Slaney estuary.

The building structure is in-situ concrete with heavyweight concrete soffits exposed throughout the public and office areas to gain benefit from the high thermal mass properties of concrete. Within the public concourse the exposed concrete soffits, in combination with the predominantly stone clad walls and floors, create a sterile environment that supported the fire strategy since the concourse functions as a principle fire escape route from the individual department blocks. Exposed soffits were selected within this space architecturally to give weight and gravitas to the spatial experience. It was also important to create a strong sense of permanence in keeping with it's function as an important and dignified civic space.

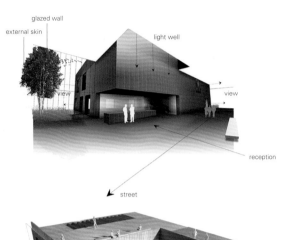

项目名称：Wexford County Council Headquarters
地点：Wexford County, Ireland
建筑师：Robin Lee Architecture
执行建筑师：Arthur Gibney and Partners　多学科工程师：Buro Happold
质量监督：Mulcahy McDonagh and Partners
景观建筑师：Mitchell & Associates
承包商：Pierce Contractors, Wexford County Council　甲方：Wexford County Council
用地面积：24,079m²　总建筑面积：11,500m²　造价：GBP 36 million
设计时间：2008　施工时间：2008.2~2011.7　竣工时间：2011
摄影师：©Andrew Lee (courtesy of the architect)

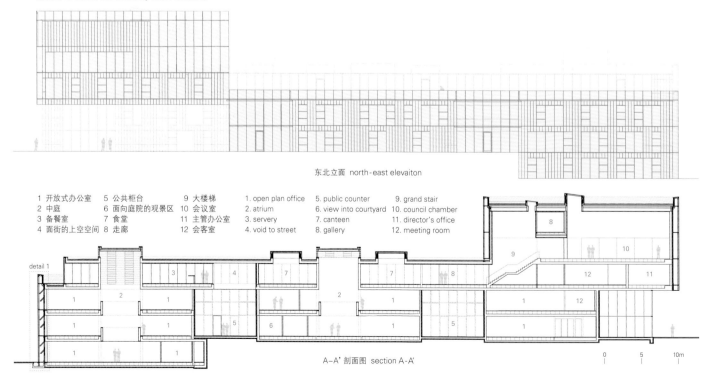

东北立面 north-east elevaiton

1 开放式办公室	5 公共柜台	9 大楼梯	1. open plan office	5. public counter	9. grand stair
2 中庭	6 面向庭院的观景区	10 会议室	2. atrium	6. view into courtyard	10. council chamber
3 备餐室	7 食堂	11 主管办公室	3. servery	7. canteen	11. director's office
4 面街的上空空间	8 走廊	12 会客室	4. void to street	8. gallery	12. meeting room

A-A' 剖面图　section A-A'

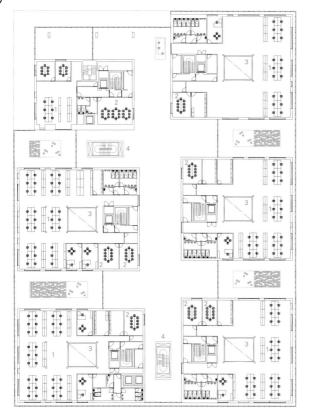

1 办公室 2 会客室 3 中庭 4 公共楼梯
1. office 2. meeting room 3. atrium 4. public stair

一层 second floor

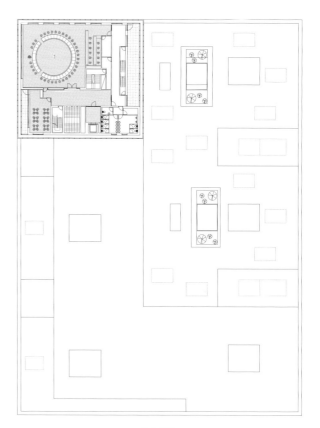

1 会议室
1. council chamber

四层 fourth floor

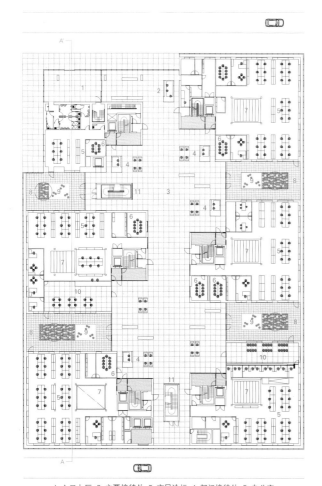

1 入口大厅 2 主要接待处 3 市民论坛 4 部门接待处 5 办公室
6 会客室 7 中庭 8 庭院 9 水池 10 公共柜台 11 公共楼梯
1. entrance hall 2. main reception 3. civic forum 4. department reception 5. office
6. meeting room 7. atrium 8. courtyard 9. pool 10. public counter 11. public stair

一层 first floor

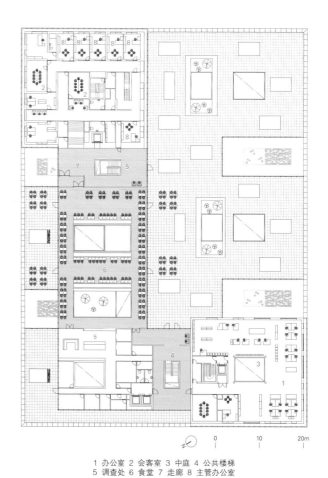

1 办公室 2 会客室 3 中庭 4 公共楼梯
5 调查处 6 食堂 7 走廊 8 主管办公室
1. office 2. meeting room 3. atrium 4. public stair
5. survery 6. canteen 7. gallery 8. director's office

三层 third floor

毕尔巴鄂市政厅
IMB Arquitectos

毕尔巴鄂市政厅的新总部位于圣奥古斯汀，临近新巴洛克宫的后立面处，新巴洛克宫建于1892年，由Joaquin Rucoba建筑师事务所设计，曾经被用作市政厅。

这一近期建成的建筑容纳了市政厅的技术办公室。而新巴洛克宫这一历史建筑则继续代表市政厅的形象，同时也还是市长办公的总部。

公共住宅空间的处理手法旨在疏散城市肌理，增加可用区域的面积，以建造一座小型广场，使其成为进入市政厅总部的城市接待处或者大厅。

位于两个体量中的分区有意将建筑融入到城市中，采用适合场地的规模和高度，且突出横跨场地的传统人行道。

而在建筑内部，建筑师曾有意在不同的楼层创造灵活的使用条例，使其能够形成不同类型的办公室，这一范围从封闭的办公室到开放式的工作区，或者是休闲的公共区。

从环保的角度来看，这座建筑应用了不同的战略，采取了不同的行动来提高工作区域的质量，同时使能源消耗和二氧化碳排放达到最小化。

周围的立面也较为智能地解决了办公室的隔离要求。

建筑南向的表皮为双层立面，内部带有维修和通风用的摄像头以及普通的软百叶条，它能够对工作站的太阳直射光以及热量吸收起到调节的作用。人工照明系统根据接收的太阳光，能够对不同区域工作台的亮度进行调节。

该设施的设计采用了最大的能效标准，并且设置了节水设备以及污水回收系统。

Bilbao City Hall

The new headquarters of Bilbao City Hall in San Agustín, is situated adjacent to the back facade of the Neobaroque Palace currently being used as town hall, built in 1892 by the project of Joaquin Rucoba Architect.

The recent building contains the technical offices of the town hall. The historic building will keep the representative image of the city corporation and the headquarters of the mayor.

The treatment applied to the residual public spaces has the objective to fluff the urban fabric and increase the available area to

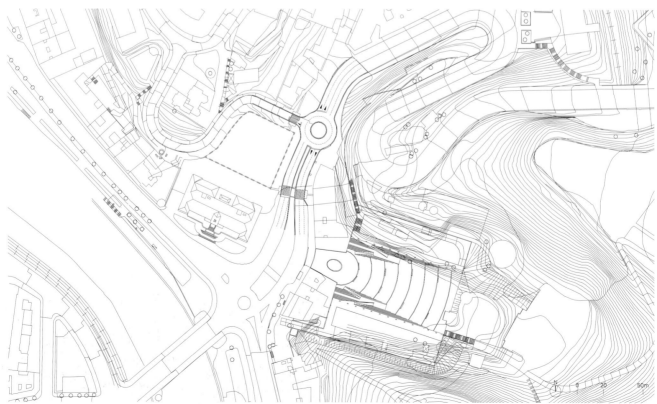

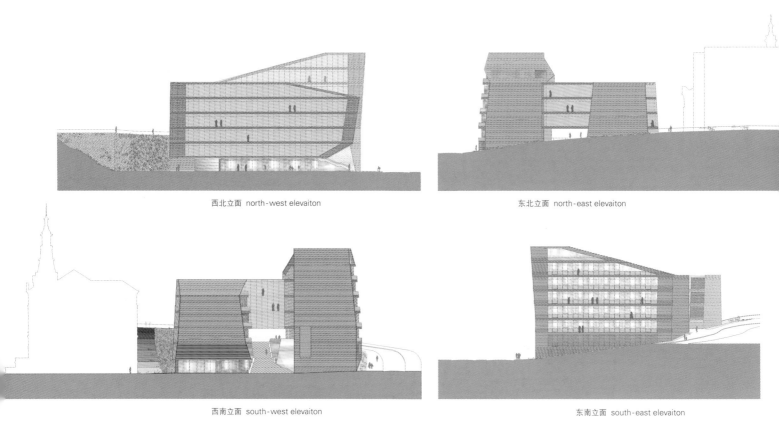

西北立面 north-west elevaiton

东北立面 north-east elevaiton

西南立面 south-west elevaiton

东南立面 south-east elevaiton

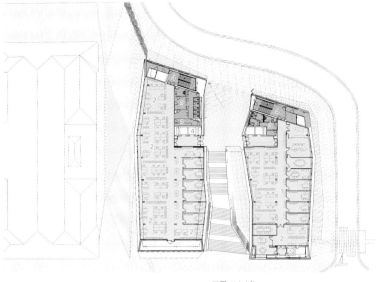

三层 third floor

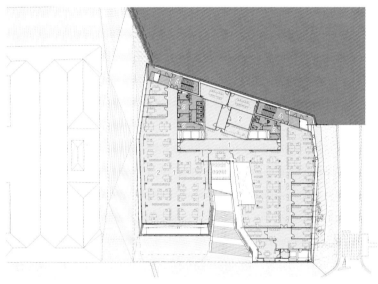

二层 second floor

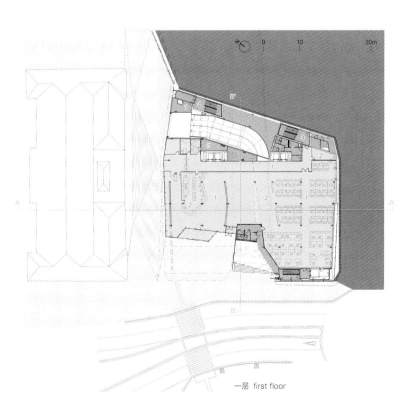

一层 first floor

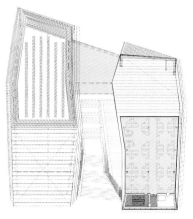

七层 seventh floor

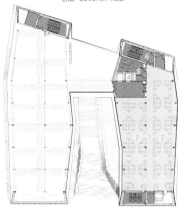

六层 sixth floor

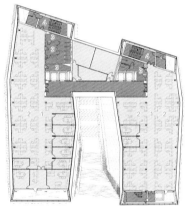

五层 fifth floor

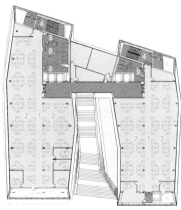

四层 fourth floor

1 大厅和交通流线区　　1. hall and circulations
2 工作区　　　　　　　2. working area
3 技术和服务间　　　　3. technical and service room
4 垂直连通区域　　　　4. vertical communications

generate a little plaza becoming an urban antechamber or lobby to access the town hall headquarters.

The fragmentation of the whole in two volumes has the will to integrate the building into the city plot, adapt the scale and the heights to the place and enhance the traditional pedestrian way across the parcel.

Inside the building the aim has been to create flexible terms of use on different floors, which are able to generate different types of offices, from the closed one to open work areas or common elements to relax.

From the point of view of the environmental improvement it has adopted different strategies and actions to increase the quality of the work area and minimize energy consumptions and CO_2 emissions.

The facade surround offers an intelligent answer to the insolation request.

The skin facing south materialized in a double facade, with maintenance and ventilation camera inside with domotic adjustable slat blinds that regulate the impact of the direct sun light on the workstations and the uptake of heat. The artificial lighting systems regulate discriminating areas the luminance above the worktop, depending the received natural light.

The facilities has been designed with maximum efficiency standards and has incorporated water consumption savings and gray water recovery.

项目名称：Bilbao City Hall
地点：Plaza Erkoreka, Bilbao, Spain
建筑师：Gloria Iriarte, Eduardo Mugica, Agustin De La Brena
合作商：Architects_Berndt Nischt, Iker Gandarias, Almudena Fernandez, Leyre de Lecea, Anartz Ormaza, Maite Eizagirre, Technical_Iban Gonzalez, Ana Ruiz, Jose Luis Castellanos
结构顾问：NB 35 SA
设备顾问：Ingeniería NIPSA SA
承包商：UTE Exbasa Amenabar
发起人：Ayuntamiento de Bilbao
有效楼层面积：13,200m²
造价：EUR 20,000,000
设计时间：2008
竣工时间：2011
摄影师：
Courtesy of the architect-p.144middle, p.144bottom, p.149
©Iñigo Bujedo Aguirre-p.142~143, p.144top, p.145, p.146, p.150, p.151

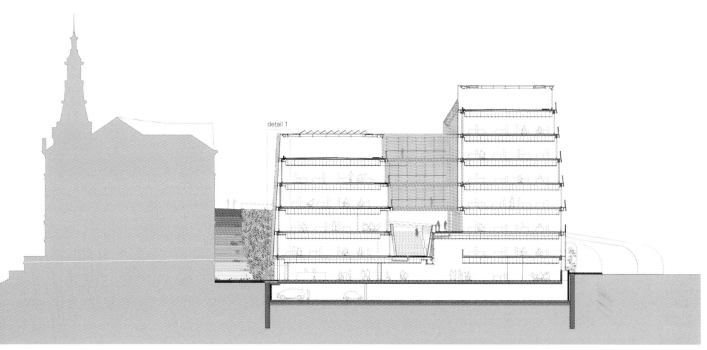

A-A' 剖面图 section A-A'

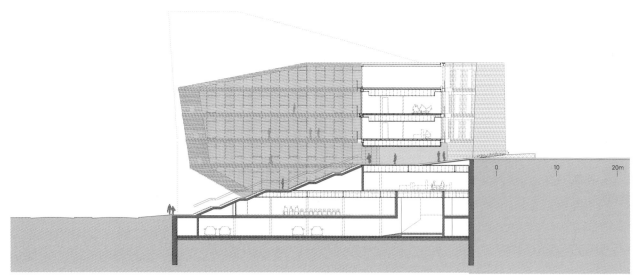

B-B' 剖面图 section B-B'

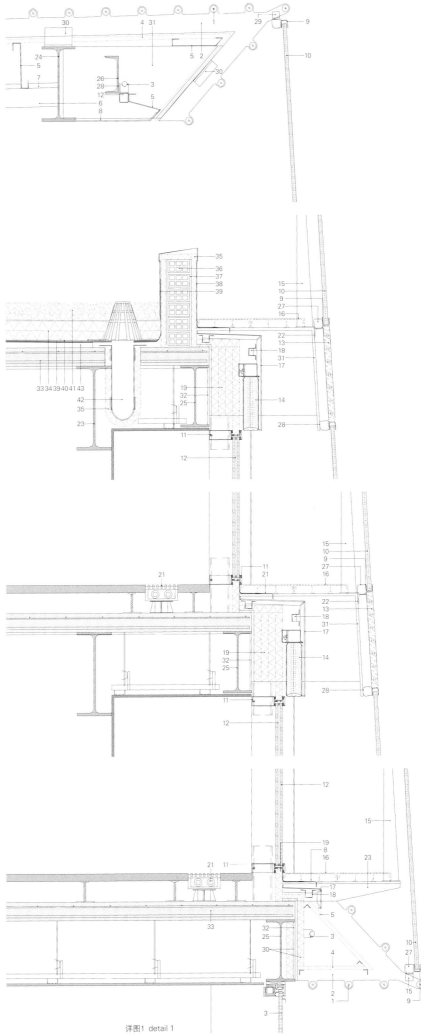

1. tubo ceramico a modo de celosia terrart baguette de NBK de seccion cirecular diametro exterios 45mm. longitud 3m. acabado esmaltado
2. subestructura metalica soporte de tubo CON tornillo M6 Y Y pasador tipo DIN 7346 3x20mm
3. luminaria estanca fluorescente IP45
4. lamina acrilica reforzada CON fibra de vidrio tipo acrylit, stabilit e: 2mm
5. omega de apoto en acero galvanizado e: 2mm+perfil
6. perfil auxiliar tipo pladur en acero galvanizado de 70mm
7. place de carton-yeso hidrofugo e: 12.5mm
8. chapa de remate plegada acero galvanizado e: 2mm
9. carpinteria en acero galvanizado MOD. FW50+especial-AWS65 de schuco o similar
10. vidrio laminar de seguridad extraclaro 8+8mm
11. muro cortina modelo FW 50 D+K de schuco o similar en perfil de aluminio anodizado
12. vidrio aislamiento tipo AN62 6+12+(4+4)mm, CON cantos pulidos
13. lamas de aluminio en forma de "Z" sobre bastidor de aluminio
14. lamas de alunimio anodizado venecianas motorizadas IP54
15. barandilla formada por pasamanos seccion circular deametro 50mm, en acero galvanizado en caliente, llanta certical 1200x50x5mm, Y sistema de sujeccion del vidrio, herraje
16. pasarela de aluminio anodizado extruido Y fresado tipo "F" alcan extruded
17. panel composite aluminio e: 4mm, tipo alipolit/FR
18. perfil omega de aluminio 40.25
19. panel de poliestireno extruido [XPS] tipo styrofoam wallmate SL.A. e: 40mm
20. viertaguas de chapa de acero galvanizado e: 2mm
21. rejilla radiante junto a hoja interior de vidrio de fachada
22. acristalamiento en planta baja laminar extraclaro 8+8mm Y sistema de sujeccion herraje tipo manet de dorma
23. perfil laminado en acero galvanizado IPE 450
24. perfil laminado en acero galvanizado IPE 400
25. perfil laminado en acero galvanizado IPE 330
26. perfil laminado en acero galvanizado UPN 200
27. perfil tubular #50.5 en acero galvanizado
28. perfil "L" 60.5 en acero galvanizado
29. perfil "L" 40.5 en acero galvanizado
30. perfil laminado en acero galvanizado "LD" 150.75.9
31. orza de acero galvanizado en caliente ancho variable e: 12.5mm
32. llanta de acero galvanizado en caliente 300x100x10mm, para anclaje de orza a estructura
33. forjado de chapa colaborante e: 100mm
34. panel de poliestireno extrudio [XPS] tipo styrofoam roofmate e: 50mm
35. panel de poliestireno extrudio [XPS] tipo styrofoam roofmate e: 30mm
36. ladrillo hueco doble
37. raseo de mortero de cemento e: 10mm
38. chapa de acero galvanizado e: 2mm, en formacion de albardilla
39. amina asfaltica de betun elastomero tipo SBS glasdan 40 P elast
40. membrana geotextil danofelt py 200 gr/m2
41. canto rodado labado diametro 25~40mm, color blanco e: 80mm
42. sumidero PVC CON bote sifonico
43. hormigon de pendiente aligerado CON arcilla expandida tipo arlita
44. loseta de hormigon prefabricado doble capa tipo lurgain 600x600x60mm

详图1 detail 1

萨莫拉郡议会建筑
G+F Arquitectos

萨莫拉郡议会建筑不仅仅给予一个特殊的规划,还有萨莫拉城市中最复杂的一处区域一个答复:Viriato广场,城市路线中一个重要的站点。这项提议为广场提供了第四个立面,使由Encarnación医院(北部)、Ramos Carrión剧院以及Condes de Alba and Aliste宫(南部)组成的建筑群更加完善。新办公室建筑围绕着一个庭院而建,并且提供了如下设施:

——复制南北朝向场地的数量,减少无自然光照射的区域。

——获取场地的对角线视角,使其周围环境成为室内生活、路线的新部分,与其他部分融为一体。

——产生一幅与建筑其他部分相对应的紧凑图景,但是也能意识到自然照明对于办公室的重要性。

立面和庭院的室外材料采用了萨莫拉生产的Arenisca石材,与历史建筑的材料一致。石材成为能够对流通风的立面,不仅仅是隐藏在结构系统中那么简单,它还通过展示其内部钢结构来突出其特点。最底层设有一个石基座,展示了城市中周围墙体使用了同样的历史语言。水平平台材质为石材,成为另一处立面。

建筑内设有两种窗户,清晰地面向室外。第一种具有一个垂直的设计理念,应用在面向广场和街道的立面中。而第二种则旨在安装在某些特定空间中,以形成合适的照明,并且应用了大量的玻璃表面。而安装在室内庭院中时,它的目地则是避免与紧凑的周围环境形成强烈的对比。

然而,建筑开放了其视野,通过穿过庭院的深洞来与广场进行连接。人们从一层的办公室进入其中,一层的办公室被用作一处郡议会公用的代表性阳台。

庭院的规模被大幅度地扩大,以允许阳光进去其周围的空间内。此外,地面层和一层属于两个不同的楼层,使望向庭院的视野来自于不同的高度,还允许不同的工作空间中留有一些距离。庭院被设计为石质表面,种有四季常青的植被。树种为杨树,纵向排列,人们在每层都能望得见。因为杨树属于落叶乔木,因此它会在每个季节都改变形象,即使在冬季,这里也会有阳光。被覆盖的入口空间的两个入口,即主入口以及独立的游客入口,十分突出。

考虑到室内空间,庭院成为空间的"规划者",以形成大型南北向分区,这些分区内的开放型办公室容纳了大量的工人,而两个小型南北办公室则用于办公和会客。封闭的办公室嵌入到北立面中,与开放的工作空间相连,来容纳需要大量隐私的员工,同时它还未与其他区域失去联系。室内布局的多功能性使建筑能够适应未来的变化。

Zamora County Council

Zamora County Council must give an answer not only to a specific program but also to one of the most complex places in the city of Zamora: Viriato Square, an important stop in the city route. The proposal offers the fourth facade to the square, completing the architectonic group of the Encarnación Hospital (North), Ramos Carrión Theater and the Condes de Alba and Aliste Palace (South).
The new offices building is organized around a courtyard that offers:

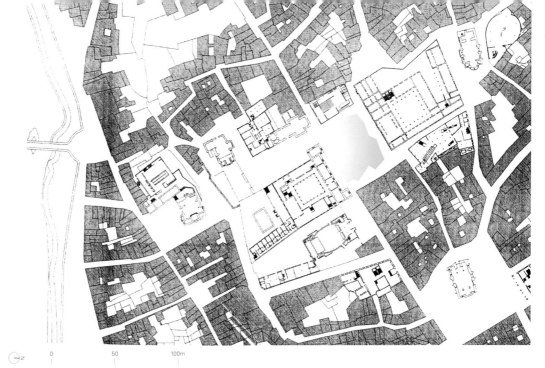

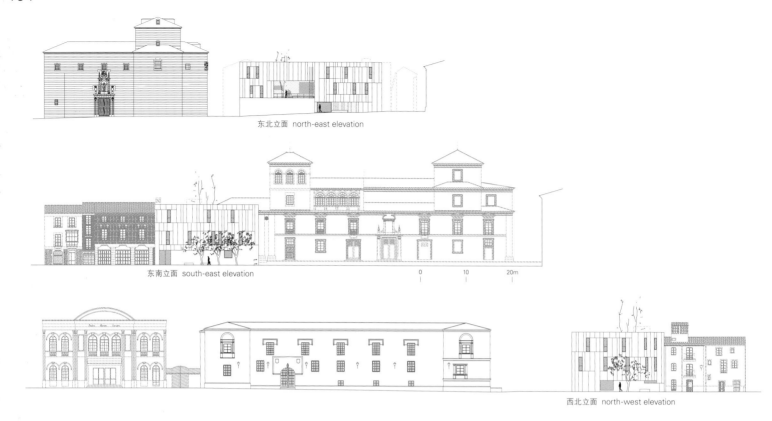

东北立面 north-east elevation

东南立面 south-east elevation   0 10 20m

西北立面 north-west elevation

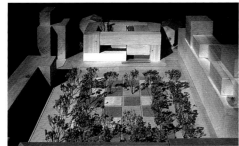

— Duplicate the number of places with North-South orientation and reduce places with no natural light.
— Obtain diagonal views to the square and its surrounding becomes new part of its interior life, its routes and its rest.
— Generate a compact image corresponding to the rest of buildings, but being conscious of necessities of natural lighting to offices.

External material of both the facade and the courtyard is solved with Arenisca stone of Zamora in the same way as the historical buildings. This stone is used as a transventilated facade that, far from hiding its constructive system, reinforces its character by means of showing its internal steel structure. At the lowest part, there is a stone base that suggests the same historical language of surrounding walls in the city. Horizontal deck made with stone is another facade.

There are two kinds of windows giving a clear face to exterior. The first ones have a vertical conception and are used in those facades to the square and streets. On the other hand, with the intention of having a correct illumination of certain spaces, large surfaces of glass are used. Being in the internal courtyard, the goal was avoiding strong contrasts with compact surroundings.

Nevertheless, the building opens its vision and contacts with the square by means of a deep hole through the courtyard. People can enter to it from offices on the first floor, being used as a representative balcony for public uses of the County Council.

The size of the courtyard has increased in high in order to allow light and sun enter into the spaces that are around it. Besides, ground floor and first floor are on two different levels so that views to courtyard aren't made from the same height and allow some distances between different working spaces. The courtyard is designed to have stone surfaces and vegetation that will be green all year. The tree is a populus alba that, for being so longitudinal, can be seen from every floor. As it is a deciduous tree, it will modify the general image in every season allowing sun in winter.

Covered access space emphasizes two entrances: the main one and the tourist office that has an independent running.

Concerning to the interior spaces, the courtyard is the organizer of space, creating larger pieces to north and south where open offices place larger number of workers, and two smaller ones to west and east where rooms with the use of offices and meetings are placed. Closed offices are inserted in north facade in connection with open working space in order to place those workers that need larger privacy in their work, but without losing proximity with the rest. Versatility in internal distribution will allow the building to adapt to future changes.

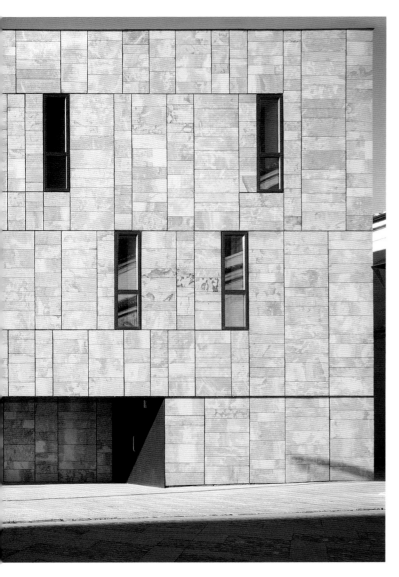

项目名称：Offices for County Council in Zamora
地点：Viriato Square, Zamora
建筑师：Pilar Peña Tarancón, María Antonia Fernández Nieto
参与竞赛的合作建筑师：Gonzalo Bárcenas Medina
合作商：Jesús García Herrero, Rafael Valín Alcocer, Jesús Hernández Alonso, Francisco Martín Gil
顾问：José María García del Monte, proyectos MYC, Jorge Gallego Sánchez-Torija
施工单位：UTE CYN-YAÑEZ/ REARASA
发起人：County Council of Zamora
用地面积：636m²
总建筑面积：576m²
有效楼层面积：2,802m²
设计时间：2002
施工时间：2008.2~2011.1
摄影师：
©Miguel de Guzmán-p.154~155, p.156, p.157, p.158
©Joaquin Mosquera-p.152~153, p.159

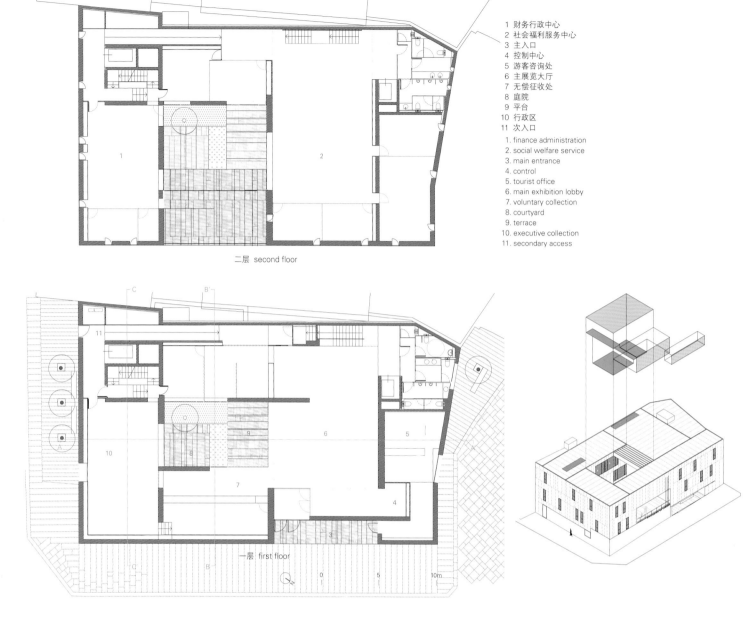

1 财务行政中心
2 社会福利服务中心
3 主入口
4 控制中心
5 游客咨询处
6 主展览大厅
7 无偿征收处
8 庭院
9 平台
10 行政区
11 次入口

1. finance administration
2. social welfare service
3. main entrance
4. control
5. tourist office
6. main exhibition lobby
7. voluntary collection
8. courtyard
9. terrace
10. executive collection
11. secondary access

二层 second floor

一层 first floor

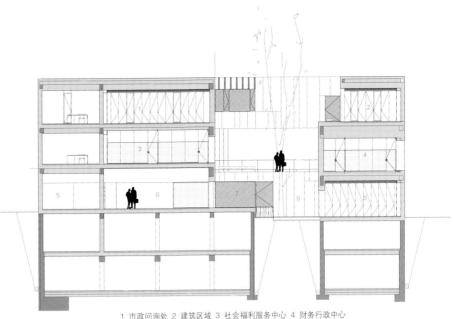

1 市政问询处 2 建筑区域 3 社会福利服务中心 4 财务行政中心
5 游客咨询处 6 主展览大厅 7 平台 8 庭院 9 行政区
1. assistance to municipalities 2. architecture area 3. social welfare service 4. finance administration
5. tourist office 6. main exhibition lobby 7. terrace 8. courtyard 9. executive collection

A-A' 剖面图 section A-A'

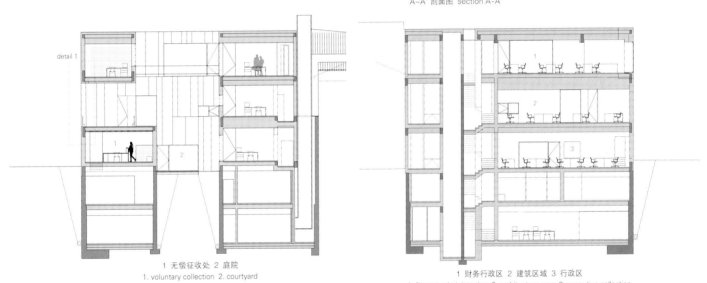

1 无偿征收处 2 庭院
1. voluntary collection 2. courtyard

B-B' 剖面图 section B-B'

1 财务行政区 2 建筑区域 3 行政区
1. finance administration 2. architecture area 3. executive collection

C-C' 剖面图 section C-C'

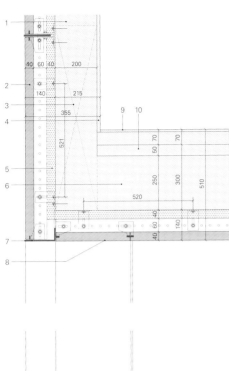
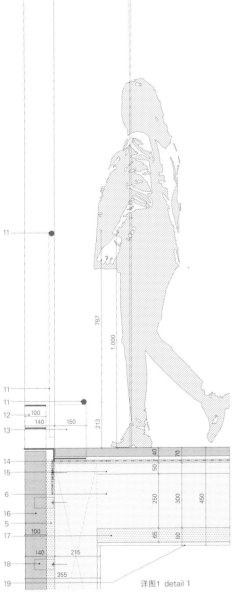

详图1 detail 1

1. precast concrete bricks, 15cm
2. quintanar sandstone, 4cm fixed with metallic structure
3. reinforced concrete structure
4. plaster board, 1.5cm
5. spray polyurethane insulation, 4cm
6. prefabricated concrete hollow-core slab, 25cm
7. edge steel angle
8. sandstone false ceiling, 4cm
9. continuous polymer flooring, 7cm
10. compression layer, 5cm
11. painted circular steel tube, d=25mm
12. steel plate sign, 8mm
13. steel plate, 250x150x20mm welded to steel plate base, 150x150x5mm
14. DPM
15. compression layer, 5cm
16. sandstone, 10cm
17. rigid insulation fixed to floor structure, 5.5cm
18. steel fixation of stone, 6mm
19. plaster board, 1.5cm

莱希河畔兰茨贝格市政厅
Bembé Dellinger Architekten

位于莱希河畔兰茨贝格的古老市政厅是由Dominikus Zimmermann于1765年建成的。

该市发起了一个小型的邀请性竞赛,以寻找扩建的列管建筑的解决方案。最原始的会议室十分小,且老建筑也缺乏残疾人通道和第二逃生路线。Bembé Dellinger建筑事务所的设计被采用。建筑师从老建筑的中心移走了游客咨询处,将其安置在北部的拱顶之下,因此能够重新开放通向后庭院的拱道。

建筑师在二层规划了一个新型会议室,使一层尽可能的整洁。一个新型入口大厅和餐饮区域在古老的历史城镇和重新恢复活力的废弃土地中创造了一个全新的、吸引人们眼球的连接。扩建的部分完全采用轻质的铜网围护起来,与原有的巴洛克式建筑相得益彰。过滤的光线有助于在会议室内传递着一种庄严的气氛。

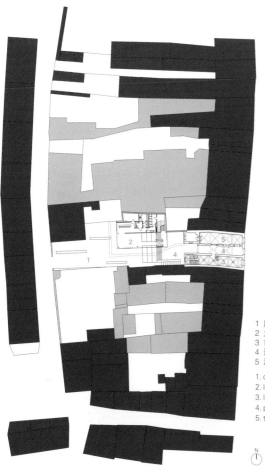

1 庭院
2 大厅
3 市政厅大厅
4 通道
5 游客咨询处

1. courtyard
2. lobby
3. lobby town hall
4. passway
5. tourist office

Landsberg am Lech Town Hall

The historical town hall in Landsberg am Lech was built by Dominikus Zimmermann in 1765.

A small invited competition was initiated to look for a solution to extend the listed building. The original council chamber had become too small, and the old building lacked access for disabled people and a second escape route. The architects' design was accepted. They moved the tourist office from the center of the old building into the northern vaults and were therefore able to re-open the ancient archway to the rear courtyard.

They planned the new council chamber on the first floor and kept the ground floor as clear as possible. A new entrance hall and dining area have created a new attractive link in the historical old town and revitalized derelict land. Totally enveloped with a light mesh of copper, the extension harmonizes well with the original Baroque building. Filtered light helps to convey a dignified atmosphere in the council chamber. Bembé Dellinger Architekten

1. steel beam, HEB 140
2. thermaly separated anchorage, schöck KST 22
3. slab with casted cavity, 32cm
4. roller blind d=ca 155mm
5. suspended ceiling, fumed oak
6. aluminium mullion-transom-facade
7. copper rib mesh on substructure
8. guide bar for roller blind
9. aluminium-grating, anodized, fixed at facade anchor
10. facade anchor, flat steel 130/30
11. aluminium sheeting, anodized, insulated with mineral wool, 20mm ventilation
12. wooden angle element, fumed oak
13. parquet, fumed oak
14. cobiax flat slab
15. aluminium faced panel with mineral wool insulation, 20mm ventilation
16. concrete piller, 30cm
17. slit gutter

详图1 detail 1

©Christoph Rehbach(courtesy of the architet)

南立面 south elevation

西立面 west elevation

项目名称: Town Hall in Landsberg am Lech
建筑师: Bembé Dellinger Architekten
合作建筑师: Architekturbüro Edenhofer-Gerum, Landsberg (site management)
承包商: City Landsberg
地点: Landsberg am Lech, Germany
用地面积: 780m²
总建筑面积: 608m²
竣工时间: 2009
摄影师: ©Stephan Müller-Naumann (courtesy of the architect) - p.160~161, p.162, p.164 (except as noted)

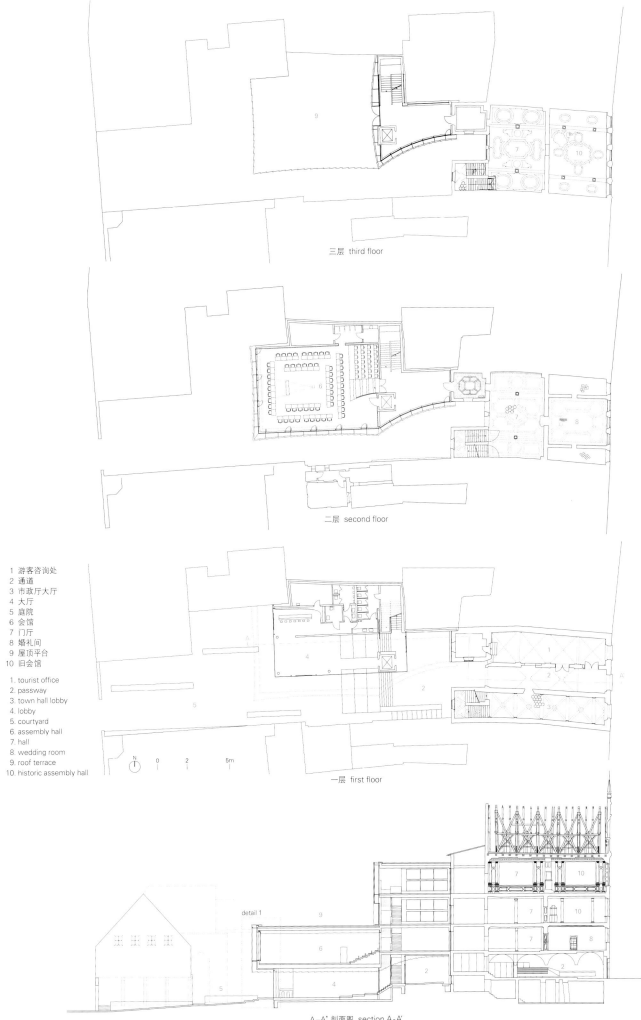

三层 third floor

二层 second floor

1 游客咨询处
2 通道
3 市政厅大厅
4 大厅
5 庭院
6 会馆
7 门厅
8 婚礼间
9 屋顶平台
10 旧会馆

1. tourist office
2. passway
3. town hall lobby
4. lobby
5. courtyard
6. assembly hall
7. hall
8. wedding room
9. roof terrace
10. historic assembly hall

一层 first floor

A-A' 剖面图 section A-A'

巴埃萨市政厅的修复
Viar Estudio Arquitectura

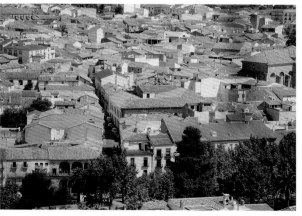

巴埃萨市政厅的修复项目曾经被解读为一段时间内存在的一个单元,以及一个不断变化的过程,而在这个过程中,新设计被认为是一个附加层,它也被解读为建筑适时所产生的最后一层沉积层。

历史建筑是以覆盖层为基础的,积累了很多不同的历史,它们可以统称为建筑的使用寿命(durée)。Henri Bergson认为建筑最根本的实质不是其本身,也不是其变化的本身,而是持续的、他称之为"durée"或者是持续时间的连续变化过程。建筑自有一种适时的存在方法,一种持续的变化过程,这种变化对于其本身来说是根本的。

这种持续时间以及连续变化的节奏意味着一种分解过程,即消减、增加、突变或者是用途改变,所有的建筑变化在岁月的流逝中都会发生。

巴埃萨市政厅项目在"durée"的建筑理念中交织着。其设计考虑了场地的额外条件,其不断的变化成为建筑的实质,以及建筑的一部分特色。

而这种概念与记忆混合的状态是建筑师将目标作为一个连续体以及关系代码的促成因素。因此,当建筑师在思考、设计以及塑造人类的历史时(这也是一种持续过程),且历史在建筑内刻上了印记,那么建筑因此便成为一种铭刻时间的方式。建筑师对每个操控对象(物质性或者是思维性的)的印象都将他们置于一处适时的地方,因为当他们建造、堆积、黏合或者是倾倒时,他们也改变了具有地质性、工业性和诗意氛围的时间。当建筑师着重刻画位于其间的重要时刻时,那么这处场地便会变得人性化,为建筑师所有,且被赋予一种人性的气息。

那么,建筑师如何理解历史建筑,便成为重要的问题。

答案缓慢地出现:建筑师将建筑比作一个不完整的部分(或者是一个树桩),一个将自身包裹起来的元素,无法展示、创造和定义其自身的结构。这一策略意图理清建筑的多余部分,以一个未完成的部分的身份去接受历史建筑,并且利用新结构将其围起来。历史建筑(即不完整的部分)并没有产生新的建筑,它是这座城市的条理,能够将现有的部分围合起来;它与城市一同生长,是拥抱城市后庭院的有机结构。

建筑分为三个部分。现存的结构:大楼,或者说历史部分,包含具有代表性和政治性的部分;新建结构:容纳办公室、公共和用户服务区;上空体量:容纳一处全新的公共空间、大厅、守候室和休息区域。

办公室建筑的楼梯是新建结构最重要的单元。在概念中,它是一个由天窗(北部和南部)照亮的巴洛克风格楼梯。仿佛人们闭上眼睛,就能够想象出大型西班牙教堂内的窗格和砌石围墙,它们将清晰的隔离闸门与金黄色的覆层、振动以及光影效果混合起来。

Baeza Town Hall Rehabilitation

Baeza Town Hall Rehabilitation project has been read as a unit in a duration, and a constant change process where the new design has been thought as an additional stratum, as the last sediment layer in time the building has created. The thought about the temporal process of architecture is fundamental.

Historical architecture is based on overlays, accumulating many different pasts in what could be called the "durée" of architecture. Henri Bergson said that the ultimate reality is not the being, nor the changing being, but the continuous process of change which he called "durée" or duration. Architecture has a way of being in time, a becoming that lasts, a change that is substance on its own. The rythm of the duration and of the successive changes connotes a dissolution process, subtraction, addition, mutation or a change of uses that befalls all architectural ensembles through time.

The Baeza Town Hall project is entwined within the concept of architectural "durée". It is designed thinking about the additive condition of the site, in the quality of change as the substance of the project and as a part of the character of the building.

The mixed state of perception/memory is what makes the architects see objects as a continuum, as relationship nodes. Thus, when they think, design or build people's memory – which is also

西南立面_原始建筑 Renaissance bay
south-west elevation_original building

西南立面
south-west elevation

东北立面_原始建筑 Renaissance bay
north-east elevation_original building

东北立面
north-east elevation

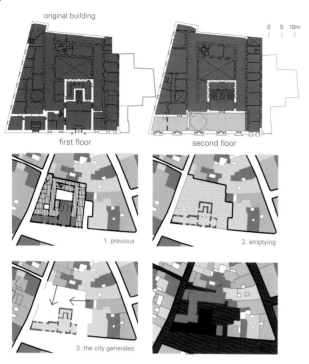

original building

first floor | second floor

0 5 10m

1. previous
2. emptying
3. the city generales

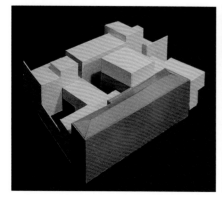 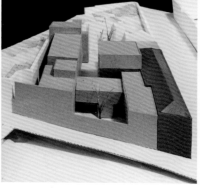 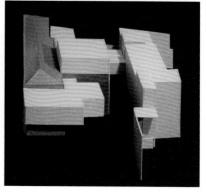

duration and imprinted in the objects and architecture becomes a way of inscribing time on matter. Man's impression on every manipulated object – material or speculative – sets the architects in a place in time because as they build, pile, glue or pour they change the geologic, industrial or poetic time of matter humanizing it, making it the architects', giving it – as they impress their vital time in it – a human breath.

The fundamental question: How do they understand the historic building?

The answer rose slowly; they think of the building as a fragment almost a stump, as an element enwrapped in itself, with no ability to suggest, nor to create, nor to define its own structure. The strategy was to clean up the building's additions, to accept the historic building as an unfinished fragment and to envelop it with new construction. The historical building, the fragment, does not create a new building; it is the town's logic which generates, encloses and wraps the existing fragment; it is the spontaneous city growth, the organic structure of its patios what hugs it.

There are three parts; the existing: the building or historic fragment which contains the representative and political part, the new: the container of offices, common and customer service areas, the void: the container of a new public space, lobby, waiting and sitting areas.

The office's staircase is the most important unit of the new part of the building. In concept it is a "Baroque" staircase illuminated by skylights (north and south). It is as if people closed their eyes and could envision the lattices and the closures of the masonry in the great Spanish churches and cathedrals which mix the clear closings of the gates with the golden coverings, the vibrations, the lights and the shadows. Viar Estudio Arquitectura

项目名称：Rehabilitation of the Town Hall of Baeza, Jaen, Spain
地点：Baeza, Jaen, Spain
建筑师：Iñigo de Viar Fraile
开发商：Junta de Andalucía, Empresa Pública de Suelo de Andalucía, Ayuntamiento de Baeza
技术指导：Iñigo de Viar Fraile, Jesús y José Luis Martín Clabo Architects
结构工程师：Minteguia y Bilbao S.L 工程师：ENER Ingenieros
建筑工程师：Luis Enrique Tajuelo Sánchez
总承包商：DRAGADOS S.A.
建筑面积：3,539m² 造价：EUR 6,481,437 竣工时间：2011
摄影师：©Fernando Alda

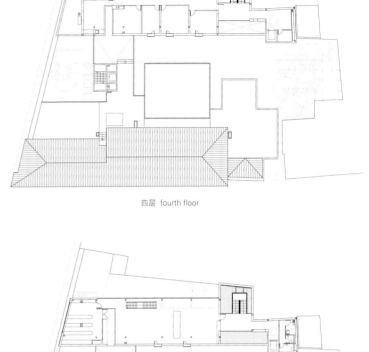
四层 fourth floor

三层 third floor

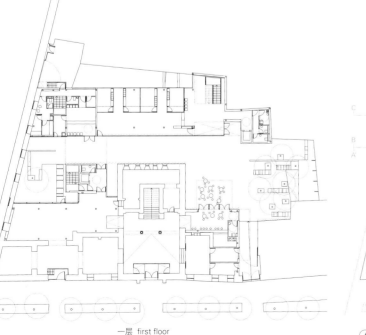
一层 first floor

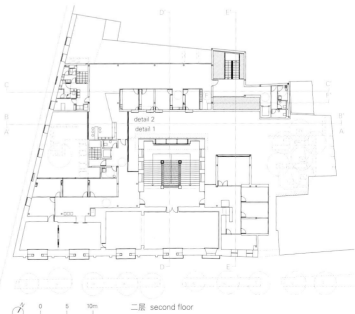
二层 second floor

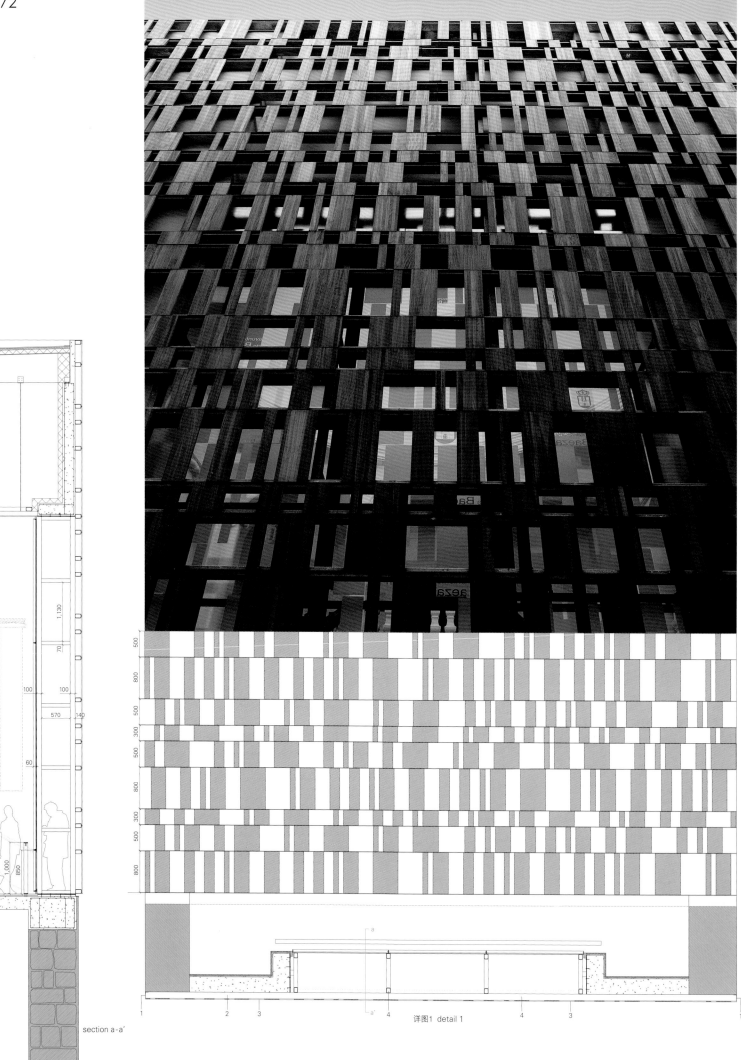

section a-a'

详图1 detail 1

173

detail 1
1. doble acristalamiento aislante con estratificado de seguridad y vidrio de baja
2. perfil hueco 100x80x5mm
3. angular calibrado acero mate pavonado inoxidable 65x50x5mm
   junquillo acero mate 2x1cm
4. perfil hueco 120x80x5mm

detail 2
1. caja chapa acero 2mm mate pavonado
2. doble perfil horizontal LD 30x20x5
3. piezas de madera natural espesor 2cm (castaño)
4. perno de anclaje
5. fadbica de bloque ceramico de arcilla aligerada de baja densidad

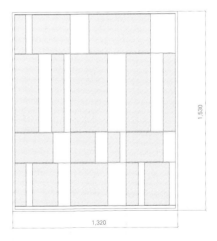

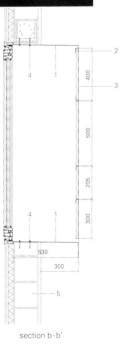

section b-b'

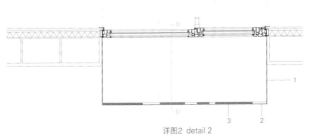

详图2 detail 2

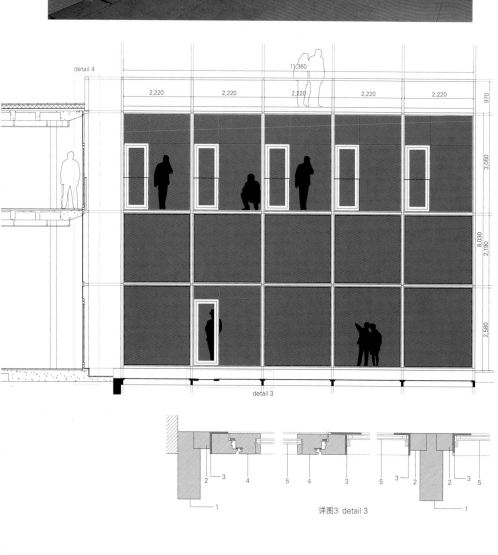
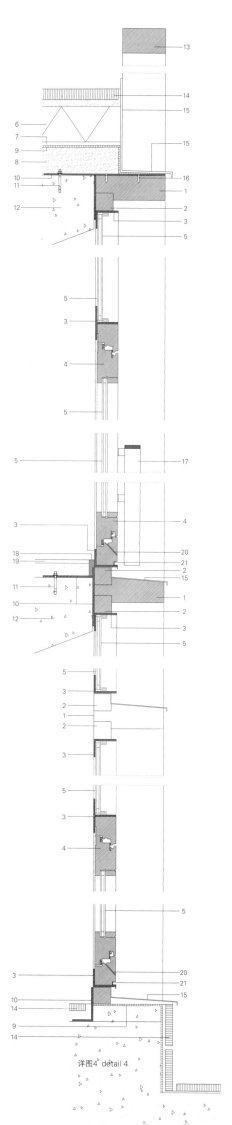

详图3 detail 3

详图4 detail 4

1. pieza macera castano 200x70mm tratada a poro abierto interior y exteriormente con aceites minerales puros y ceras de alta pureza medicinal, sin disolventes
2. pieza madera castaño 50x50mm tratada a poro abierto interior y exteriormente con aceites minerales puros y ceras de alta pureza medicinal, sin disolventes
3. angular calibrado acero mate pavonado inoxidable 65x50x5mm
4. corpinteria registable madera castano tratada a poro abierto con aceite minerale puros y ceras de alta pureza medicinal, con herrajas y mecanismo de acero mate pavonado
5. doble acristalamiento aislante con estrativicado de seguridad y vidrio de baja emisividad al interior 6+10+6 (de seguridad en planta baja)
6. planchas rigidas de aislamiento termico de poliestireno extuido para cubiertas 100mm
7. fieltro geotextil
8. capa nivelacion hormigon de arido ligero (2% pendiente) y raseo de mortero do cemento
9. lamina impermeabilizante de caucho sintetico (EPDM)
10. chapa acero 10mm union carpinteria-forjado losa armada definido en plano herreria
11. anclaje metalico de seguridad por socabado automatico, version esparrago
12. forjado losa armada 30cm
13. barandilla madera castano tratada a poro abierto interior y exteriormente
14. baldosa de piedra natral sobre cama de mortero 20mm
15. chapa acero inoxidable mate 1mm goteron
16. tornillo union chapa y madera
17. barandilla pletina calibrada 45x10mm
18. junta elastica
19. pavimento en goma sobre capa de nivelacion
20. rebosaderos en carpinteria de madera
21. goteron do chapa galvanizada doblada e: 1mm atornillada a la carpinteria

1. forjado reticular de hormigón armado de 300mm
2. forjado reticular de hormigón armado de 350mm
3. zanca de hormigó armado con formación de escalones
4. bloque cerámico de arcilla aligerada de baja densidad 19cm
5. ladrillo jueco doble 30x14x6cm
6. mortero de agarre
7. malla de unión de materiales
8. mortero de nivelación 4cm
9. enfoscado de bortero de cemento hidrófugo
10. rastrel de madera
11. mortero de cal color marfil
12. peldaño de madera natural de wengé 290x30cm
13. contrahuella de madera natural de wengé 170x20cm
14. tablero nadera encolada en espesores variables con base de pintura rojo burdeos y acabado en pan de oro
15. piezas de cuelgue del falso techo
16. perfil matálico para sujeción de falso techo
17. tablero estratificado de madera natural impergnada en resinas para formación de falso techo y revestimiento vertical. color wengé
18. perfil para fijación de placas de yeso laminado
19. panel placa de yeso laminado 13mm
20. panel placa de yeso laminado 15mm
21. perfil laminado L180
22. perfil conformado en frío LDF 40x20x3
23. perfil hueco laminado 200x200x10mm
24. perfil IPE 180mm
25. perfil laminado L65
26. perfil hueco cuadrado 50x50x4mm
27. pletina de acero calibrado de 5mm en "T" para sujeción de los montantes de madera
28. escalera metálica. perfil hueco 160x100mm
29. pieza metálica para sujeción de peldaño
30. barandilla de acero galvanizado, según detalle en plano
31. angular calibrado acero mate pavonado inoxidable 65x50x5mm
32. carpintería de aluminio
33. anclaje metálico de chapa versión tornillo ó espárrago para cargas medias, longitud 100mm. mín 70mm de perforación
34. pieza de borde de forjado bloque ceramico de baja densidad 9.6m
35. tornillo de lata resistencia
36. pieza de madera natural de castaño oscuro 60x30mm
37. montante de madera natural de castaño oscuro 60x30mm
38. aislamoento térmico rígido de poliestireno extruido para muros
39. aislamoento térmico rígido de poliestireno extruido para cubiertas
40. lámina impermeabilizante de caucho sintético (EPDM)
41. lucernario conformado por vidrio de control solar y baja emisividad y vidrio templado de alta resistencia 6+10+4+4
42. cubierta de zinc: soporte de madera, aislamiento rígido tipo roofmate 100mm, soporte de madera para fijación de Memvrana delta WMZ y láminaexterior de zinc de 0.8mm de espesor
43. teja cerámica plana de gran formato (TBF)
44. tablero hidrófugo para soporte de cubierta
45. muro medianero
46. ipunto de luz directa ø15cm o equivalente
47. pantalla suspendida du luz directa e indirecta, fabricada en aluminio y policarbonato 4x24W 698x698mm
48. pieza madera castaño 35x65mm tratada a poro abierto con aceites minerales puros, añadiendole
49. pigmentado film de PVC o polietileno

Archidona的文化中心和市政厅

Ramón Fernández-Alonso Borrajo

从行政的角度来说，这个项目可以被称之为Archidona附属区旧学校广场的修复建设，项目位于Ochavada广场以及San José街之间，将两种完全不同类型的嵌入结构结合在一起，一方面作为一座面向广场的遗产建筑，主要对其功能进行重新定义，且进一步突出其结构特色。同时，新建筑与San José街相邻，且建造方法较为独特。

该项目将要求的功能区进行排列，使其与现有的历史建筑的空间承载力兼容，同时还为功能区的其他部分规划了一处新体量，这个体量（文化功能）的规模要求更大。为了达成这一目标，建筑师提出了两种类型的嵌入结构，一方面，为了尽可能地覆盖这座18世纪的建筑，建筑师致力于在场地设置具有代表性的房间以及市政厅，另一方面，新建部分用来容纳文化功能区：大厅、展览大厅、一层的餐厅以及上层的办公室。新体量依附于历史建筑，是一个雕刻玻璃式的棱柱体，其布局和材料的使用都较为独立，玻璃幕墙的设计也用来控制光线，以解决新老建筑之间的过渡问题，模糊各立面之间相交线。

Cultural Center and the New City Hall of Archidona

The project, called administratively "Rehabilitation of the Old School Square Minor Archidona" located in the Ochavada Plaza and the San José Street, combines two distinct types of intervention, consisting on one hand, in acting on the heritage building that gives face to the square consisting primarily of a conceptual redefinition of its functioning and enhancer of its structural features and, at the same time, the approach of a new building facing the street's frontage of the San Jose Street with its own specificity.

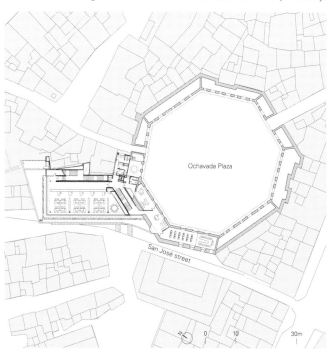

项目名称：Cultural Centre and the New City Hall of Archidona
地点：Archidona, Andalusia, Spain
建筑师：Ramón Fernández-Alonso Borrajo
质量监督：Miguel Ángel Jiménez Dengra, Rafael Palma Moyano
结构工程师：Ramón Fernández-Alonso y Asociados SL.
机械工程师：Miguel Ángel Martínez Lebrusant Ingenieros S.L.
施工单位：Constructora San José S.A
家具制造：Decofisur. Juan José Fernández
照明工程师：ALS Lighting. Antón Amann
土地开发商：Junta de Andalucía, Ayuntamiento de Archidona
用地面积：925m² 总建筑面积：2,864m² 有效楼层面积：2,221m²
造价：EUR 4,441,133.97
设计时间：2007 竣工时间：2010
摄影师：©Fernando Alda

The project orders the requested program, making it compatible with existing space capabilities into the historic building and also proposes a new volume for those parts of the program that requires larger (cultural program). To this end the architects propose two types of interventions, on one side, to recover as much as possible the building of the XVIII, they located the areas dedicated to the representative room and the city hall, and on other hand on the new part to accommodate the program cultural: hall, exhibition hall, restaurant on the ground floor and offices on upper floors. Attached to the historic building, this new volume is a prism carving glass of marked independence in configuration and materials, in which a "curtain of glass" is designed to control the light, solve the transition between the old and new, obscuring the line of intersection of their respective planes of the facade.

Ramón Fernández-Alonso Borrajo

东南立面 south-east elevation

东北立面 north-east elevation

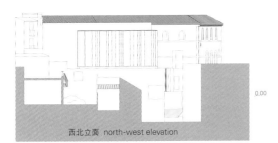
西北立面 north-west elevation

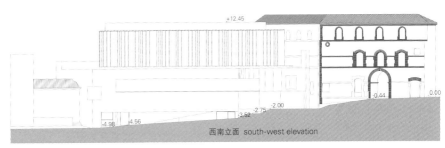
西南立面 south-west elevation

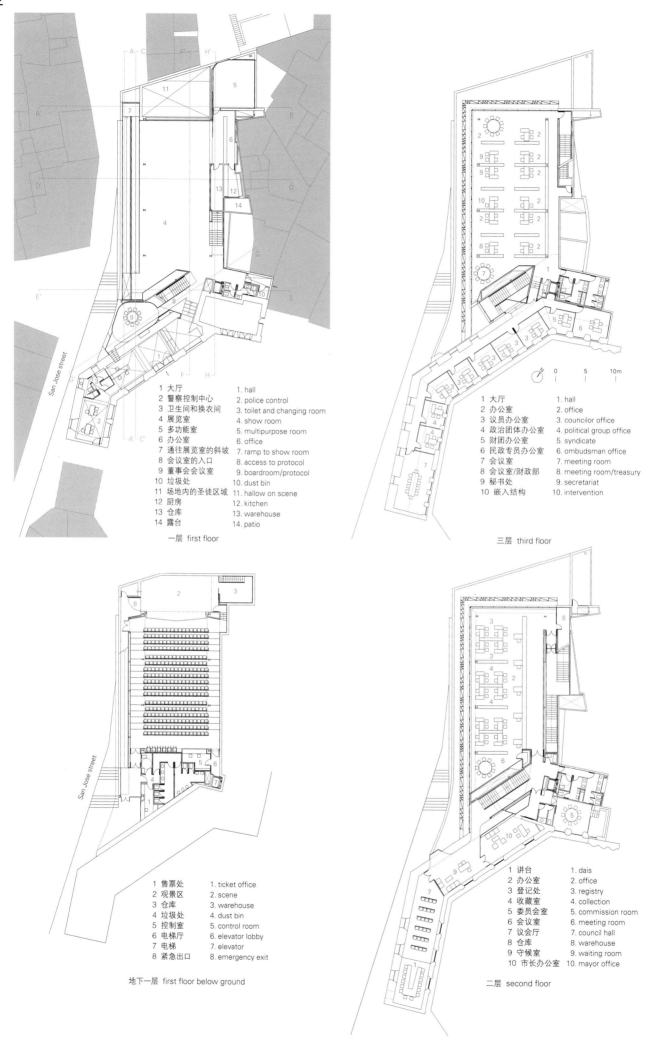

1. canto rodado, ø=16/32mm, espesor min, 50mm
2. anclaje de fachada progección solar con IPN 100
3. remate superior de biombo con pletina de acero inoxldable, e=4mm
4. UPN 350
5. vidrio extraciaro 12+12 stabip/butiral opal
6. boble stadip 6+6 butiral transparente +cámare de aire 12mm. en carpinteria fija de aluminio tipo laminex serie A45, anodizado en su color
7. tramex galvanizedo 45x45mm
8. perfil IPN 80 para sujeción de pasarela
9. mantenimiento chapa alucodond e. 4mm
10. falso techo de cartón-yeso, con carpintería galvanizada y aislamiento térmico
11. pavimento téchico(cánara 10cm)
12. placas de cartón-yeso sobre perfilería metálica
13. perfil "U" para recepción de vidrio

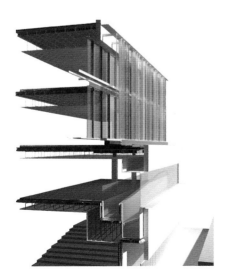

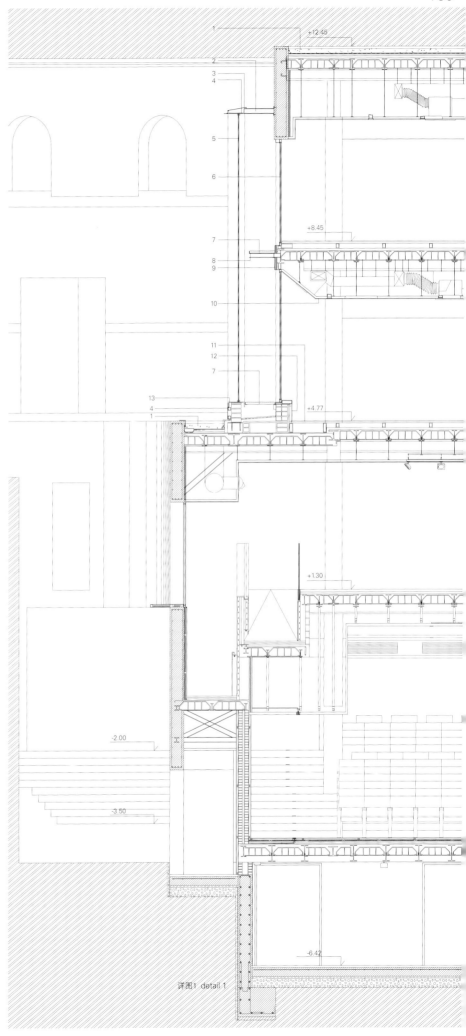

详图1 detail 1

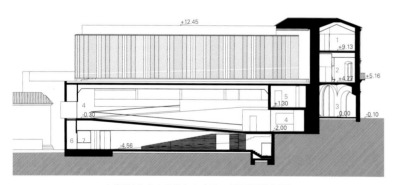

1 议员办公室 2 守候室 3 大厅 4 通往展览馆的斜坡
5 董事会会议室 6 垃圾处 7 紧急出口
1. councilor office 2. waiting room 3. hall 4. ramp to salon
5. boardroom 6. dust bin 7. emergency exit
A-A' 剖面图 section A-A'

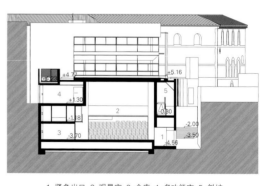

1 紧急出口 2 观景室 3 仓库 4 多功能室 5 斜坡
1. emergency exit 2. scene 3. warehouse 4. multipurpose room 5. ramp
B-B' 剖面图 section B-B'

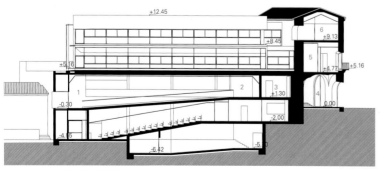

1 通往展览室的斜坡 2 展览室 3 董事会会议室 4 大厅 5 守候室 6 议员办公室
1. ramp to show room 2. show room 3. boardroom protocol 4. hall 5. waiting room 6. councilor office
C-C' 剖面图 section C-C'

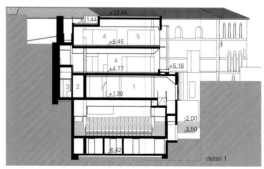

1 展览室 2 仓库 3 厨房 4 办公室 5 秘书处
1. show room 2. warehouse 3. kitchen 4. office 5. secretariat
D-D' 剖面图 section D-D'

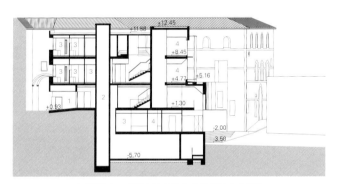

1 通往Salazar街的入口 2 电梯 3 卫生间 4 会议室
1. access to Salazar Street 2. elevator 3. toilet 4. meeting room

E-E' 剖面图 section E-E'

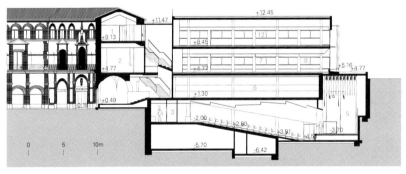

1 议员办公室 2 市长办公室 3 卫生间 4 大厅
5 观景室 6 展览室 7 收藏室 8 登记处 9 办公室
1. councilor office 2. mayor office 3. toilet 4. hall
5. scene 6. show room 7. collection 8. registry 9. office

F-F' 剖面图 section F-F'

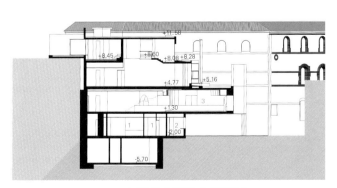

1 卫生间 2 售票处 3 会议室入口
1. toilet 2. ticket office 3. access to protocol room

G-G' 剖面图 section G-G'

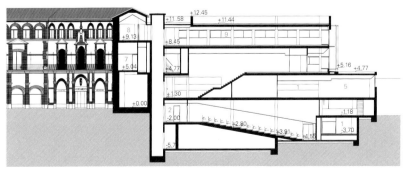

1 仓库 2 电梯厅 3 多功能室 4 卫生间 5 财团办公室 6 办公室
1. warehouse 2. elevator lobby 3. multipurpose room 4. toilet 5. syndicate office 6. office

H-H' 剖面图 section H-H'

Bembé Dellinger Architekten
Was established in 1997 by Felix Bembé and Sebastian Dellinger. Felix was born in Munich, Germany in 1969. Received a diploma from Munich University of Applied Sciences in 1994 and established Bembé Dellinger Architekten in 1997. Sebastian was also born in Munich, Germany in 1966. Received a diploma from Munich University of Applied Sciences in 1996.

>>72

Hierve
Is a boutique design consultancy based in Mexico City and London established in 1999 with purpose of serving society through high quality projects in architecture, product design, visual arts and service design. Has architecture philosophy finding a balance between the following four factors; creativity, function, social responsibility, and spirituality. Their work have been published in several art magazines and exhibited in mostly Mexico.

>>124

Carreño Sartori Arquitectos
Was founded in 2000 by Mario Carreño Zunino and Piera Sartori del Campo in Santiago, Chile. Mario Carreño Zunino graduated from Pontifical Catholic University of Chile(PUCCH) in 2000. He has been lecturing at the architecture school of the PUCCH since 2003. Piera Sartori del Campo also graduated from the PUCCH in 1999 and studied landscape at the PUCCH till 2003. She has been teaching at the architecture school of the PUCCH since 2003. They are willing to link architecture to territorial area, connecting materials nature to different constructive systems. Their work has been part of various publications, exhibitions and biennials.

>>30

Ateliers Jean Nouvel
Jean Nouvel was born in Fumel, France in 1945. Studied at the Ecole des Beaux-Arts in Paris. Has headed his own architectural practice since 1970. Established Ateliers Jean Nouvel in 1994. Was awarded the Pritzker Prize in 2008. His major completed buildings are Arab world Institute in Paris Lyon Opera House, Lucerne Culture and Congress Center and etc.

>>30

Habiter Autrement
Mia Hägg was educated at Chalmers University of Technology in Gothenburg and National School of Architecture of Paris-Belleville. From 1998 to 2001, she was employed by Ateliers Jean Nouvel. Her collaboration with Herzog & de Meuron began in 2002. In 2003, she was selected as the project manager for the National Stadium of the 2008 Olympic Games in Beijing, and she became an associate of the firm in 2005. In 2007, she left Herzog & de Meuron to found her own practice, Habiter Autrement.

>>134

Robin Lee Architecture
Is an architectural firm established in Glasgow in 2002. Has offices in London and Dublin.
Robin has a background in architecture and sculpture, having supplemented his full range of architectural qualifications with a postgraduate diploma in fine art (sculpture) at Glasgow School of Art in 1993. Their projects are informed by a rigorous interrogation of context through research and analysis and by a process of open dialogue with clients, users and consultants.

>>166

Viar Estudio Arquitectura
Iñigo de Viar finished his studies in architecture at the School of Architecture of San Sebastian in 1988. Has been a professor of projects since 1989 at the School of Architecture of San Sebastian. Has completed projects such as Boulevard of Leioa and Planetarium of Pamplona in Spain, project for the seafront, multi-purpose building and parking on the beach in Conil de la Frontera (Cádiz), Town Hall of Conil de la Frontera, Baeza Town Hall Rehabilitation, and project of the Cultural Center for Youth in Zarautz, Spain.

>>94

LEM+ Architectes
Pierre Lépinay and Bertrand Meurice met in 2000 while working in an agency. In 2004, they decided to join forces and submitted a competition entry for the multimedia library in Ballan-Miré, France. In the same year, they created their own agency which has recently been renamed LEM+ Architectes. Their projects covers a wide range of programs including housing, cultural amenities, schools and business premises. The main goal is to allow the occupants to appropriate and personalize the space. They offer distinctive buildings whose forms directly result from their specific settings. Their projects seek to make best use of modern technologies while also taking full advantage of climatic resources.

>>84

Agence Bernard Bühler
Bernard Bühler was born in 1954. Spent his childhood in Bordeaux Chartrons dock. At the age of 18, he received B.E.P and was trained himself in practical business for several years. Became a member of French Institute of Architects in 1985. Established his own architectural firm in 1986 and has been collaborating with other offices in Chartrons since then.

>>142

IMB Arquitectos
Is a consultant architects firm for the development of architectural projects, urban records, advice and works site direction. Was founded in Bilbao, Spain by Eduardo Mugica van Herckenrode[middle], Gloria Iriarte Campo[right] and Agustin de la Brena Torres[left] after a continuous professional collaboration since 1979. Three of them were all born in Bilbao. Eduardo graduated from the University of Madrid in 1979 and Gloria graduated from the University of the Basque Country in 1979. Augustin graduated from the University of the Basque Country in 1983.

>>108

Aedes Studio
Is a design collective based in Sofia, Bulgaria. Was set up in 2003 by Plamen Bratkov[right] and Rossitza Bratkova[left]. They have received several architectural

>>152

G+F Arquitectos

María Antonia Fernández Nieto[left] and Pilar Peña Tarancón[right] were born in 1972 and studied architecture in ETSAM. Have collaborated in the development of several projects from 1998 to 2003. In their broad experience, they have received many prizes from the competitions such as 22 Social Housing in Torrelodones, The New Main Square of Belmonte, and Elder Residence building in Higueruela. Their work has been exhibited in El Croquis Gallery Space in EL Escorial, Madrid and has been published in BAU magazine, Arquitectura Viva, Arquitectos COAM.

Jorge Alberto Mejia Hernández

Had his education as an Architect at the Universidad del Valle and graduated in 1996. Holds a Master in History and Theory of Art and Architecture (2002) as well as a Master degree in Architecture (2008), both from the Universidad Nacional de Colombia. His teaching includes Architectural Theory, History of Architecture and Design Studio at the Universidad Nacional de Colombia since February 2005, where he became Professor Catedratico Asociado in 2007.

Has written many books such as of Enrique Triana: Obras y Proyectos (Bogotá: Planeta, 2006) and Coauthor of Vivienda Moderna en Colombia (Bogotá: Universidad Nacional de Colombia, 2004) and XX Bienal Colombiana de Arquitectura (Bogotá: Sociedad Colombiana de Arquitectos, 2006). His research interests include architectural form, modern architecture, contemporary conditions and architectural principles and procedures.

Aldo Vanini

Practices in the fields of architecture and planning. Had many of his works published in various qualified international magazines. Is a member of regional and local government boards, involved in architectural and planning researches. One of his most important research interests is the conversion of abandoned mining sites in Sardini

>>100

Cohlmeyer Architecture Limited

Is an architectural firm established in 1981. Provides architectural services for a wide range of projects including apartments, schools, offices, hotels, commercial and institutional buildings. Their design work has been honored in national and international publications and design awards. Steve Cohlmeyer has been working in the profession since 1968, and has been president of Cohlmeyer Architecture since 1981. He studied architecture at Harvard University. Is a fellow of the Royal Architectural Institute of Canada. Has had a parallel career as a fine artist. Has had exhibitions in public and private art galleries.

>>100

5468796 Architecture

Is a Winnipeg based architecture collaborative with expansive interests engaging in all aspects of design including visioning, branding, architecture, object design, detailing and engineering systems. The office was formed in 2007 and continually seeks to challenge convention in a multitude of scales. The studio constantly dissects design and culture through a series of ongoing as a manifestation of individual preference, but rather as a direct response to client, context, and program.

>>178
Ramón Fernández-Alonso Borrajo
Received his degree in architecture from the School of Architecture in Madrid in 1981. Has been a professor at the University of Granada(UGR) since 1997. Also taught at the University of Navarra(UNAV).

>>38
DAO
Was established by David Arias Aldonza[left] and Cristina del Buey García[right]. They are in searching of connection of ideas that can interact with reality, transforming it and providing valid and long-lasting content. David was trained over a period of 8 years in the architectural office of AMP, Fernando Menis. and worked in an engineering office for 2 years in Madrid. Cristina worked with Andrés Celis for 2 years in Burgos. Next step was in an engineering office in Madrid for 5 years.

>>46
Alfredo Payá Benedito
Was born in Alicante, Spain, in 1961. Received a master's degree from Technical Superior School of Architecture of Madrid, Spain in 1990. Received a Ph.D in architecture, University of Madrid, in 2005. Has organized seminars and workshops and diverse architectural meetings. His complete work has been reviewed in different publications and national and international exhibitions. At the beginning of 2006, he founded his own architectural office noname 29, where he develops his work till present day.

>>56
Jaques Moussafir Architectes
Jaques Moussafir was born in 1957. Studied at the Ecole d'Architecture de Pairs-Tolbiac and art history at Sorbonne University with Daniel Arasse. Established his firm in 1993 after training over a period of 10 years in the architectural office of Christian Hauvette, Henri Gaudin, Dominique Perrault and Francis Soler. Gave lectures at a number of European schools such as Polytechnic University of Barcelona and the school of architecture in Valle. Has been continuing lecturing as associate lecturer at several universities in Paris since 2003.

>>64
KOZ Architectes
Was founded in 1999 by Christophe Ouhayoun[left] and Nicolas Ziesel[right]. Both graduated from the Paris-Belleville School of Architecture. KOZ creates non-typical, friendly and sensitive buildings that redefine public spaces and emphasize the diversity of their uses. Champions architecture that is aware of context and creates surprise and "added value" in terms of function, by making use of residual areas that are easily adapted to stimulate the imagination of local residents, users and visitors. Regularly carries out wide ranging assignments including project management, site management, auditing, signage, and furniture work. The team is made up of 8 to 12 persons architects, computer graphics designers, liaison officer and manager. Everyone is involved in information coordination and participation in decision-making throughout the project.

C3, Issue 2013.8

All Rights Reserved. Authorized translation from the Korean-English language edition published by C3 Publishing Co., Seoul.

© 2013大连理工大学出版社
著作权合同登记06-2013年第267号

版权所有·侵权必究

图书在版编目(CIP)数据

居住的流变：汉英对照 / 韩国C3出版公社编；张琳娜，于风军，耿婷婷译. —大连：大连理工大学出版社，2013.11
（C3建筑立场系列丛书）
Dwelling Shift
ISBN 978-7-5611-8328-1

Ⅰ. ①居… Ⅱ. ①韩… ②张… ③于… ④耿… Ⅲ. ①住宅－建筑设计－汉、英 Ⅳ. ①TU241

中国版本图书馆CIP数据核字(2013)第270705号

出版发行：大连理工大学出版社
　　　　（地址：大连市软件园路80号　邮编：116023）
印　　刷：北京雅昌彩色印刷有限公司
幅面尺寸：225mm×300mm
印　　张：12
出版时间：2013年11月第1版
印刷时间：2013年11月第1次印刷
出 版 人：金英伟
统　　筹：房　磊
责任编辑：张昕焱
封面设计：王志峰
责任校对：高　文

书　　号：ISBN 978-7-5611-8328-1
定　　价：228.00元

发　行：0411-84708842
传　真：0411-84701466
E-mail：12282980@qq.com
URL: http://www.dutp.cn